Susanne Kamptmann

**REACH Compliance –
The Great Challenge for
Globally Acting Enterprises**

Related Titles

Schecter, A. (ed.)

Dioxins and Health

Including Other Persistent Organic Pollutants and Endocrine Disruptors, Third Edition

2012
ISBN: 978-0-470-60529-5
(Also available in digital formats)

Deutsche Forschungsgemeinschaft (DFG) (ed.)

List of MAK and BAT Values 2011

Maximum Concentrations and Biological Tolerance Values at the Workplace. Report 47

2011
ISBN: 978-3-527-33061-4

Sterner, O.

Chemistry, Health and Environment

Second Edition

2010
ISBN: 978-3-527-32582-5

Lewis, R.J.

Hazardous Chemicals Desk Reference, Sixth Edition

Sixth Edition

2008
ISBN: 978-0-470-18024-2 (Also available in digital formats)

Susanne Kamptmann

REACH Compliance – The Great Challenge for Globally Acting Enterprises

Verlag GmbH & Co. KGaA

The Author

Dr. Susanne Kamptmann
Am Weiheracker 5
79585 Steinen
Germany

All books published by **Wiley-VCH** are carefully produced. Nevertheless, authors, editors, and publisher do not warrant the information contained in these books, including this book, to be free of errors. Readers are advised to keep in mind that statements, data, illustrations, procedural details or other items may inadvertently be inaccurate.

Library of Congress Card No.: applied for

British Library Cataloguing-in-Publication Data
A catalogue record for this book is available from the British Library.

Bibliographic information published by the Deutsche Nationalbibliothek
The Deutsche Nationalbibliothek lists this publication in the Deutsche Nationalbibliografie; detailed bibliographic data are available on the Internet at <http://dnb.d-nb.de>.

© 2014 Wiley-VCH Verlag GmbH & Co. KGaA, Boschstr. 12, 69469 Weinheim, Germany

All rights reserved (including those of translation into other languages). No part of this book may be reproduced in any form – by photoprinting, microfilm, or any other means – nor transmitted or translated into a machine language without written permission from the publishers. Registered names, trademarks, etc. used in this book, even when not specifically marked as such, are not to be considered unprotected by law.

Print ISBN: 978-3-527-33316-5
ePDF ISBN: 978-3-527-66432-0
ePub ISBN: 978-3-527-66431-3
Mobi ISBN: 978-3-527-66430-6
oBook ISBN: 978-3-527-66429-0

Cover Design   Bluesea Design, McLeese Lake, Canada
Typesetting   Toppan Best-set Premedia Limited, Hong Kong
Printing and Binding   Markono Print Media Pte Ltd, Singapore

Printed in Singapore
Printed on acid-free paper

Contents

Foreword *XIII*
Preface *XV*

1 **Introduction** *1*
1.1 History *1*
1.2 The REACH Regulation – A Short Overview on the Table of Contents *3*
1.3 Purpose and Scope of REACH *4*
1.4 Other Regulations and Directives that are Important in the Context of REACH *7*
1.4.1 Fees and Charges Payable to the European Chemicals Agency *8*
1.4.2 Competition Law *9*
1.4.3 GHS and CLP *10*
1.4.4 Other Regulations Containing the Wording REACH *11*
References *11*

2 **Roles under REACH** *15*
2.1 Manufacturer within the EU *15*
2.2 Non-EU Manufacturer, Importer and Only Representative *16*
2.3 Downstream User *20*
2.4 Trader within EU versus Non-EU Trader and Distributor *23*
2.5 Examples and Exercises *25*
References *26*

3 **What Sort of Substances have to be Considered under REACH** *27*
3.1 Substance, Mixture and Article under REACH *27*
3.2 Different Compositions *28*
3.2.1 Mono-constituent Substance *28*
3.2.2 Multi-constituent Substance *28*
3.2.3 Substances of Unknown or Variable Composition, Complex Reaction Products or Biological Materials *29*
3.3 Different Types of Use *29*

3.3.1	Substance with End Use 29
3.3.2	Intermediate 30
3.3.2.1	Non-isolated Intermediate 30
3.3.2.2	On-site Isolated Intermediate 30
3.3.2.3	Transported Isolated Intermediate 31
3.4	Phase-In Substances 31
3.5	No-Longer Polymers 32
3.6	Non-Phase-In Substances 32
3.7	Substances that Already Have Been Notified 33
3.8	Overview on Official EC Numbers and not Official List Numbers 33
3.9	Exemptions from REACH 34
3.9.1	Non-isolated Intermediates 35
3.9.2	Substances Manufactured or Imported in Amounts below 1 t/a 35
3.9.3	Substances Mentioned in Annex IV 35
3.9.4	Substances Listed in Annex V 36
3.9.5	Substances in the Interest of Defense 36
3.9.6	Waste and Recovered Substances 37
3.9.7	Polymers 37
3.9.8	Re-imported Substances 39
3.9.9	Further Exemptions: Use in Medicinal Products or for Food and Feedingstuffs 39
3.9.10	Product and Process Orientated Research and Development 40
3.9.11	Substances Regarded as Being Registered 40
3.9.12	How to Cope with Situations in Which Parts of the Manufactured Amount are Falling under REACH and Another Part is Exempted 42
3.10	Check-List for Business Managers 43
3.11	Examples and Exercises 44
	References 47

4	**Obligation to Submit a Registration Dossier** 49
4.1	Who has to Register? Who may Register? 49
4.2	Pre-registration and Late Pre-registration 53
4.3	When Does a Substance have to be Registered? 55
4.4	Special Rules for Non-EU Manufacturers 56
4.5	Consequences for Globally Acting Enterprises/What to Take into Account within a Decision-Making Process? 59
4.6	Examples and Exercises 59
	References 60

5	**Types of Registration** 61
5.1	Standard Registration, Full Registration or Registration as a Substance 70
5.2	Registration as an On-site Isolated Intermediate 70
5.3	Registration as a Transported Isolated Intermediate 71
5.4	Formerly Notified Substances 80

5.5	PPORD 80
5.6	Examples and Exercises 80
	References 81

6	**Data Requirements and Dossier Preparation** 83
6.1	Data Requirements 83
6.2	Dossier Preparation 84
6.2.1	PPORD 89
6.2.2	Inquiry Dossier 90
6.2.3	On-site Isolated Intermediate 91
6.2.4	Transported Isolated Intermediate 93
6.2.4.1	Check-List for Preparation of the Substance Data Set in IUCLID5 95
6.2.5	Standard Registration (Full Registration) 95
6.3	Some Useful Tips for Entering Data and Information in Certain Chapters in IUCLID5.4 98
6.3.1	IUCLID Section 1.2 98
6.3.2	IUCLID Section 1.3 99
6.3.3	IUCLID Section 1.4 100
6.3.4	IUCLID Section 1.7 101
6.3.5	IUCLID Section 2.3 101
6.3.6	IUCLID Section 3.1 102
6.3.7	IUCLID Chapter 11 102
6.3.8	IUCLID Chapter 13 103
6.4	Data Requirements, Type of Registration and Costs/Fees 103
6.5	Examples and Exercises 105
	References 105

7	**Claiming a Registration Number for Already Notified Substances** 107
7.1	Formerly Notified Substances are Regarded as Registered under REACH 107
7.2	How to Claim the Registration Number Under REACH for a Formerly Notified Substance 108
7.3	When to Update a Registration Dossier of a Formerly Notified Substance and How to Do It 109
7.4	Examples and Exercises 109
	References 110

8	**Process for Registration of Non-Phase-In Substances** 111
8.1	Inquiry Dossier 112
8.2	Preparation of the Registration Dossier 113
8.2.1	Registration as Member of Joint Submission 115
8.2.2	Registration within a Joint Submission in Cooperation with Other Potential Registrant(s) 115

8.2.3	Single Submission *116*	
8.3	Difficulties and Problems that can Arise in the Context of the Registration of Non-Phase-In Substances *117*	
8.4	Examples and Exercises *118*	
	References *119*	
9	**Process for Registration of Phase-In Substances** *121*	
9.1	Preparing for Pre-registration and Late Pre-registration *121*	
9.2	Communication within Pre-SIEF *124*	
9.2.1	Data Holders *125*	
9.2.2	Third Party Representatives *126*	
9.2.3	Potential Registrants *126*	
9.2.4	Duties and Rights of the Different SIEF Participants *126*	
9.3	Formation of SIEF *127*	
9.3.1	Substance Sameness and Substance Identification Profile (SIP) *130*	
9.3.2	Lead Registrant Agreement *131*	
9.3.3	Lead Registrant Notification *134*	
9.4	Cooperation within the SIEF *134*	
9.4.1	Obligations of SIEF Participants *135*	
9.5	Data Sharing *135*	
9.5.1	Consortium Agreement *137*	
9.5.2	Cooperation Agreement *137*	
9.5.3	SIEF Agreement *137*	
9.5.4	Letter of Access *137*	
9.5.4.1	Letter of Access Concerning Data as Studies and Tests *138*	
9.5.4.2	Letter of Access to a Registration Dossier *139*	
9.6	Data Sharing Disputes *140*	
9.7	Opt-Out *142*	
9.8	Registration Dossier of the Lead Company and Registration Dossiers of the Members of Joint Submission *144*	
9.9	Overview on Important Steps within the Process for Registration of Phase-In Substances *145*	
9.10	Examples and Exercises *145*	
	References *146*	
10	**What Happens after Submission of Your Registration Dossier to ECHA?** *147*	
10.1	Initial Verification *147*	
10.2	Overall Completeness Check *149*	
10.3	Receiving the Reference Number *150*	
10.4	End of Pipeline Activities *150*	
10.5	Dossier and Substance Evaluation *152*	
10.5.1	Examination of Testing Proposals *152*	
10.5.2	Compliance Check of Registration *153*	
10.5.3	Substance Evaluation *154*	

10.5.4	On-site Isolated Intermediates are not the Object of Evaluation	*155*
10.6	Further Obligations of the Registrant and Downstream Users	*155*
10.6.1	Safety Data Sheets and extended Safety Data Sheets	*155*
10.6.2	Documentation of Correspondence with Customers Purchasing Transported Isolated Intermediates	*157*
10.6.3	Substance Volume Tracking	*157*
10.6.4	Obligation to Update Information	*158*
10.6.5	Obligations of Downstream Users	*158*
10.7	Examples and Exercises	*159*
	References	*160*
11	**Update of the Registration Dossier**	*163*
11.1	When to Update Your Registration Dossier	*163*
11.2	Requested Updates	*164*
11.2.1	Update Requested Because of Missing Information	*164*
11.2.2	Updates Requested as a Result of Dossier Evaluation	*165*
11.2.2.1	Update Requested as a Result of a Compliance Check	*165*
11.2.2.2	Update Requested after Examination of Testing Proposals	*166*
11.3	Spontaneous Updates	*166*
11.3.1	Update Because of Change in Status or Identity of the Registrant	*167*
11.3.2	Update Because of Change in the Composition of the Substance	*168*
11.3.3	Update Because of Change of Tonnage Band	*169*
11.3.4	Update Because of New Identified Uses	*169*
11.3.5	Update Because of New Knowledge of the Risks of the Substance	*170*
11.3.6	Update Because of Any Change in Classification and Labeling	*170*
11.3.7	Update Because of an Amendment in the Chemical Safety Report	*171*
11.3.8	Update Because of the Need to Perform Further Tests	*171*
11.3.9	Update Because of a Change in the Access Granted to Information in the Registration	*171*
11.4	Update of Dossiers of Formerly Notified Substances	*172*
11.4.1	Update Because of Tonnage Band Increase for Former Notified Substances	*172*
11.4.2	Other Updates for Former Notified Substances	*173*
11.4.3	Confidentiality Claims that were Previously Done in the Notification	*173*
11.5	Update of Dossiers for PPORD Notifications	*174*
11.6	Costs Concerning Updates	*175*
11.7	Examples and Exercises	*176*
	References	*178*
12	**Substances of Very High Concern and Authorization Process**	*181*
12.1	Uses that are Exempted from Authorization	*181*
12.2	Substances of Very High Concern (SVHC)	*182*

12.3	Substance Identification and Identification Procedure 184
12.3.1	Identification Procedure 185
12.3.2	Content of an Annex XV Dossier 185
12.3.2.1	Annex XV Report for the Identification of a Substance as a CMR, PBT, vPvB or ELOC 187
12.4	Inclusion of a Substance in the Candidate List of Substances of Very High Concern (SVHC) 187
12.5	Prioritization and Inclusion of Certain SVHCs in Annex XIV 188
12.6	Information in Annex XIV 188
12.7	Restrictions and Information in Annex XVII 189
12.8	Application for Authorization 190
12.8.1	Main Elements of an Application for Authorization 191
12.8.1.1	Adequate Control Route 194
12.8.1.2	Socio-economic Assessment (SEA) Route 196
12.9	Data Requirements and Documents Needed for an Application for Authorization 196
12.9.1	Substance Identity and Composition Concerning IUCLID Sections 1.1 and 1.2 197
12.9.2	Identifiers to be Entered in IUCLID Section 1.3 198
12.9.3	Identification of the "Uses Applied for" Concerning IUCLID Section 3.5 198
12.9.4	Assessment Reports Concerning IUCLID Chapter 13 199
12.9.5	Information to be Provided in the Dossier Header 199
12.10	Submission of the Application of Authorization, Deadlines and Fees 200
12.11	Subsequent Applicants and Their Obligations 201
12.12	Process after Submission of the Application for Authorization 201
12.12.1	Requested Update 202
12.12.2	Spontaneous Update 202
12.12.3	To Dos after Granting of an Authorization 202
12.12.4	To Dos after Refusal of an Authorization 202
12.12.5	Review of Authorizations 203
12.13	Examples and Exercises 203
	References 204

13 Achieving REACH Compliance within Your Company – How to Implement Processes to Ensure Legal Compliance 207

13.1	List of Used Raw Materials 207
13.1.1	Define Your Role under REACH 208
13.1.2	Define the Registration Deadline Based on Properties 211
13.1.3	Identify Uses of a Certain Substance within Your Company 211
13.1.4	List of Raw Materials when Our Company is a Downstream User 214
13.1.5	Process after Receiving a SDS or an eSDS from Your Supplier 216

13.1.5.1	Four Key Steps in Checking Safety Data Sheets and Exposure Scenarios *216*	
13.1.5.2	Information to be Forwarded to Customers down the Supply Chain *216*	
13.2	List of Substances that are Manufactured in Your Company *217*	
13.2.1	Identification of Registration Obligations *218*	
13.2.2	Define the Registration Deadline for Substances Manufactured within Your Company Based on Their Properties and Consider Consequences if a Substance is Included in the SVHC Candidate List *218*	
13.2.3	Uses at the Company's Own Site and Identified Uses of the Customers *221*	
13.3	Documentation Concerning Manufacturing Process of OIIs and TIIs and Documentation of the Correct Use of TIIs by Customers *224*	
13.3.1	In-house Documentation *224*	
13.3.2	Confirmation from Downstream Users Concerning Art. 18(4) and Further Exemptions *224*	
13.4	Substance Volume Tracking *225*	
13.4.1	Substance Volume Tracking for EU Manufacturer *225*	
13.4.2	Substance Volume Tracking for a Non-EU Manufacturer *225*	
13.5	Examples and Exercises *228*	
	References *228*	

14 Communication in the Supply Chain *229*

14.1	Communication Obligations According to the REACH Regulation *229*
14.2	Communication to be Done by Suppliers *231*
14.2.1	Communication from Supplier to EU Customers *231*
14.2.2	Communication from Supplier to Non-EU Customers *232*
14.2.3	Information for Workers *233*
14.2.4	Communication with Upstream Supplier *234*
14.2.5	Communication with Authorities *234*
14.2.6	Further Communication Obligations for Suppliers in the Context of Authorisation *235*
14.3	Communication to be Done by Non-EU Manufacturers *235*
14.3.1	Communication from Non-EU Manufacturer to his Only Representative *236*
14.3.2	Communication from Non-EU Manufacturer to EU Customer *236*
14.3.3	Communication from Non-EU Manufacturer to Non-EU Customers *236*
14.3.4	Communication from Non-EU Manufacturers to Their Suppliers *241*
14.4	Communication to be Done by Non-EU Distributors or Non-EU Traders *243*
14.4.1	Communication with EU Customers *244*

14.4.2	Communication with Non-EU Manufacturers 245
14.4.3	Communication with Only Representative Acting on behalf of a Non-EU Manufacturer (Supplier) 246
14.5	Communication to be Done by a Downstream User or a Downstream Supplier 247
14.5.1	Communication with Suppliers 247
14.5.2	Communication with Only Representative of Non-EU Manufacturer 248
14.5.3	Communication with Customers (Downstream Users) 248
14.5.4	Communication from Downstream User with Workers 249
14.5.5	Communication from Downstream User to Authorities 249
14.6	Communication to be Done by an Only Representative 249
14.6.1	Communication with Non-EU Manufacturer 250
14.6.2	Communication with ECHA and National Authorities 250
14.6.3	Communication with Customers of the Non-EU Manufacturer 250
14.7	Examples and Exercises 253
	References 254

Appendix – Answers and Solutions Concerning the Sections Examples and Exercises within this Book 255

Index 281

Foreword

Considering the political and economic challenges the European Union is presently facing, one might wonder, how it came to pass, but REACH is one of the largest legislative projects of the union ever.

Obviously, the driving force behind this has been the widespread deep concern about chemical substances in the environment, in food, in articles of daily use, in cosmetics, in detergents, in toys and at the workplace. In the past, the European continent has indeed experienced the destructive consequences of frivolous or even grossly negligent if not criminal exposure to chemicals. The author of these lines experienced in his childhood whitecaps in horribly looking and smelling rivers as everyday occurrences. Furthermore, serious environmental and health disasters have happened. Since those days, much has changed for the better. The quality of the environment has improved to a degree that some of the once most-polluted rivers are now used as bathing waters. This happened as a result of a combination of a number of causes: societal pressure, that led to new laws and governmental institutions, technological improvements, the achieved prosperity, changing mentality and priorities, a relocation of manufacturing to other parts of the world and the deindustrialization of entire regions of Europe.

It is a great challenge for the lawmaker to find the right balance between further improving the protection of health and the environment on the one hand and stifling economic growth by creating bureaucratic monsters on the other. From my point of view, the legislator has not always succeeded.

Today's legislation and in particular REACH are not just a couple of clear-cut commands and interdictions to be obeyed. As I am writing these lines, the consolidated version of REACH, which was last updated in September 2012, comprises 516 pages including the annexes. At the same time, even some single paragraphs of Article 3 on definitions, which one should think is the simpler part of the legislation, have become subject to lengthy opposed legal opinions or even to lawsuits. Moreover, the legislation deals with an extremely complicated and diverse subject matter: substances, mixtures and articles. Already, the universe of chemical substances that ranges from inorganic compounds, which already in themselves form a large kingdom, to the empire of organic compounds, to gases, to the many metals, which are used by mankind. A large variety of human activities are in the scope of the legislation, ranging from domestic everyday activities and pastimes, to artisanal small trade, to commerce, to large-scale industrial manufacturing. What makes it even trickier is that very often there is sometimes not just one solution, which is compliant, but there might be a specific solution that is the best in a particular situation.

Moreover, to make it even more difficult, REACH forces the different commercial actors to interact or even to collaborate with each other either within the supply chain or horizontally between competitors. This fact introduces confidentiality issues, supplementary economic considerations and potential problems related to antitrust legislation. And all of a sudden, the personnel responsible for the compliance with REACH are confronted with complicated legal aspects or with "exotic" topics like social-economic analysis. At the same time, complex scientific and technical problems need to be resolved in the field of toxicological and ecotoxicological testing and risk assessment.

There is hardly anybody within the concerned industries who has not heard about REACH. Even outside Europe, where REACH triggered numerous regulatory initiatives around the world, REACH has become almost a generic word for chemical legislation. Sometimes, I am asked the question when will REACH eventually be over. The clear answer is: no end is scheduled. REACH is designed to evolve on and on.

It is recommended that beginners to this field receive a well-structured introduction. Before one can get involved with the above-mentioned complicated, high-level problems, a novice needs a sound understanding of the basic craft. Without having understood the detail, it is difficult to competently deal with the subject. Therefore, I recommend starting your career as a regulatory expert here, with the next page.

Basel, October 2013

Jan Backmann
Head Chemical Legislation
F.Hoffmann-La Roche Ltd

Preface

Globally acting enterprises have to face the fact that it is not enough just to offer products of good quality, but they sell directly or indirectly a large number of additional services, for example, packaging, transportation, clearance and sometimes also taking back (parts of the) product and treatment as waste.

In almost every state of the world there are laws, directives and regulations concerning different aspects that all have to be considered to be in legal compliance when dealing with chemical substances. As customers expect their suppliers to take care of all necessary actions to achieve legal compliance it is a necessary capital investment when having the aim to ensure that business will go on in the future.

For every chemical company, especially in the case of globally acting enterprises, this is a great challenge. Ensuring legal compliance is a time-consuming and also costly process. There is a need for well-trained personnel in the Regulatory Affairs Department that know what to do to achieve legal compliance and how to do it in time, but the decision-makers (Chief Executive Officers, Business Managers, Project Managers among others) have to be aware of the most important aspects of certain regulations to be in the position to make good business decisions. They have to estimate the costs for the achievement of legal compliance and take them into account when they offer the product and they have to be aware of the risks that accompany with the decisions they make.

A lot of detailed information has to be considered to make the right decisions. In industry, economic aspects are as important as legal compliance and the idea to prevent humans and environment from any harm.

As the REACH regulation and its effects on different parts of globally acting enterprises is of high complexity, it is important that decision makers understand the basics of REACH to develop more than one option how to achieve REACH compliance in their company. In a decision-making process different options can be evaluated to find advantages and disadvantages of each option.

It is the aim of this book to show the fundamental aspects of REACH in only a few pages to decision makers that may read the book either from the beginning to the end or via several chapters as they never will have the time and patience to study all the boring legal documents, but are interested in minimizing the risks of making the wrong decisions concerning REACH matters.

At the end of every chapter there are given some examples and also exercises that help to think through and understand in more depth certain aspects. It may be that there is, as in real life, more than only one solution to achieve the desired REACH compliance.

Hopefully, the case studies will encourage the readers to look for more than one option to solve the problems that arise in their own companies and afterwards to evaluate the perfect solution for their special case.

I am sure it will be rewarded at a later stage with saving time and costs for your company and you will additionally have the good feeling that you made your job at least a little bit better than you would have done without turning over the leaves of this book.

However, please be aware that the content of this book is based on my best knowledge, but as regulations and laws are living documents they can be amended or the meaning could be changed. Therefore, whenever in doubt, please check the latest version of the mentioned documents provided by the authorities or ask a competent lawyer. I am not in the position to undertake responsibility for any of your business decisions.

October 2013

Dr. Susanne Kamptmann
Steinen

1
Introduction

REACH or **REACh** is the widespread and well-known abbreviation for a European regulation that is entitled "Regulation (EC) No 1907/2006 of the European Parliament and of the Council" [1]. This regulation entered into force on 18 December 2006.

The single letters of the word REACH go back on the subtitle of this regulation [1] that tells us something about the content of this piece of legislation – **R**egistration, **E**valuation, **A**uthorization and Restriction of **Ch**emicals (REACH) [1].

1.1 History

The concept to notify substances is not completely new. Before REACH entered into force European manufacturers or importers of new substances were obliged to notify these substances with the national authorities when they intended to produce or import a new substance in amounts of 10 kg/year or more. This was demanded by the predecessor(s) of REACH [1], whereas in Directive 67/548/EEC [2] also classification, packaging and labeling of dangerous substances was regulated. Directive 67/548/EEC [2] was repealed by the Regulation (EC) No 1272/2008. The former notified new substances are called NONS (notified new substances).

Already well-known substances that had been marketed before 1981 and therefore were listed in the EINECS [3] were not considered at all.

In the past there had been national laws and rules concerning the handling of chemical compounds. EU enterprises doing business with other companies in different countries had to struggle with a jungle of not clearly arranged rules in their daily business. It was hardly possible to follow up all amendments of the different laws being in force in different countries.

Therefore, it was an aim to harmonize the laws in the member states of the European Union. As a starting point the European Commission in Brussels made directives. Every directive had to be converted into national law(s) by all member states.

Very often this resulted in smaller or greater discrepancies between certain member states, because the directive was transformed in a slightly different way

REACH Compliance – The Great Challenge for Globally Acting Enterprises, First Edition.
Susanne Kamptmann.
© 2014 Wiley-VCH Verlag GmbH & Co. KGaA. Published 2014 by Wiley-VCH Verlag GmbH & Co. KGaA.

European Economic Area (EEA)	
28 EU Member States	**plus 3 further states**
Austria	Norway
Belgium	Iceland
Bulgaria	Liechtenstein
Croatia*	
Cyprus	
Czech Republic	
Denmark	
Estonia	
Finland	
France	
Germany	
Greece	
Hungary	
Ireland	
Italy	
Latvia	
Lithuania	
Luxembourg	
Malta	
Netherlands	
Poland	
Portugal	
Romania	
Slovakia	
Slovenia	
Spain	
Sweden	
United Kingdom	

* Access of Croatia in July 2013.

Figure 1.1 European Economic Area (EEA) consisted of 30 states (June 2012).

into national laws. In some cases it was necessary to amend existing national laws, because European right has a higher priority than already existing laws in a member state.

The situation is completely different with REACH as it is a European regulation. Therefore, REACH is directly valid in the whole European Economic Area (EEA), consisting of the member states of the European Union plus Norway, Iceland and Liechtenstein (see Figure 1.1) without any necessity to transform it into national law.

As a direct consequence there had to be established a new European Authority the so-called **European Chemicals Agency (ECHA)** that is located in Helsinki (Finland) [4].

ECHA is in charge of all European manufacturers, importers and Only Representatives (ORs) that registered already or intend to register chemical substances under REACH as a prerequisite to manufacturing and selling chemical compounds within the European Union/EEA.

Completely new and maybe the most important amendment to former laws is that now industry is responsible for the protection of humans and the environment. "No data, no market" [5] is the burden industry has to bear in mind.

The fact that REACH applies not only to new chemical compounds but also to all the substances listed in EINECS is also new [3].

As mentioned before, at present REACH is applied in the European Economic Area (28 EU member states plus Norway, Iceland and Liechtenstein). Although REACH does not apply in Non-EU countries, REACH also has a great influence on globally acting enterprises outside of the EEA.

In particular, it is a great challenge for companies located in states on the European continent, but who are not a member of the European Union itself, for example, Swiss companies. However, in having a certain knowledge of REACH matters, these internationally acting enterprises sometimes can even benefit from REACH. We will see that in detail at a later stage.

1.2
The REACH Regulation – A Short Overview on the Table of Contents

The REACH regulation is a bulky document but as it is subdivided into several parts that are themselves divided up into single articles you will be in the position to have an overview of this complex topic within a short time.

After having understood the general principles you can study the relevant parts of this piece of legislation in depth whenever it should become necessary.

Within this chapter I will give only a short overview on the content of the REACH regulation as the aim of this book is to concentrate on the most important aspects.

The title of the "Regulation (EC) No 1907/2006 OF THE EUROPEAN PARLIAMENT AND OF THE COUNCIL of 18 December 2006" [1] gives in the first part the number of the regulation (No 1907) and the year when it entered into force (No 1907/2006) and it is also stated who passed the law (OF THE EUROPEAN PARLIAMENT AND OF THE COUNCIL), and finally, followed by the exact date (18 December 2006) when it entered into force.

The subtitle describes the subject of the regulation, in this case "concerning the Registration, Evaluation, Authorization and Restriction of Chemicals (REACH), establishing a European Chemicals Agency, amending Directive 1999/45/EC and repealing Council Regulation (EEC) No 793/93 and Commission Regulation (EC) No 1488/94 as well as Council Directive 76/769/EEC and Commission Directives 91/155/EEC, 93/67/EEC, 93/105/EC and 2000/21/EC" [1].

In the beginning of the regulation there is a rationale as to why it has been taken into consideration to work out this piece of legislation. There are also references to already existing directives and regulations and the kind of influence they will have on REACH or how REACH will have an influence on the already existing laws. It is a sort of evaluation of which regulation or directive will have an effect on other pieces of legislation and of which one will have a higher priority. As in

Section (15) of this statement in the beginning of the regulation the conclusion is made that it is necessary to establish a European Chemicals Agency "to ensure effective management of the technical, scientific and administrative aspects of this Regulation at Community level" [4], all registrants to cooperate with the same European Authority. This "independent central entity" is the well-known ECHA in Helsinki.

After 131 short sections of such statements as mentioned above, finally the table of contents can be found within the REACH regulation [1].

The REACH regulation consists of fifteen Titles and so far seventeen Annexes. Each title is subdivided into several chapters. Each chapter contains several articles. However, the articles of the REACH regulation are numbered consecutively and independently of the chapters (see Figure 1.2).

The most import titles are the following:

Title I deals with general issues. First, we find a chapter about "Aim, scope and application" [6], followed by "Definitions and general provision" [7].

Title II devotes to "Registration of substances" [8], whereas **Title III** deals with "Data Sharing and Avoidance of unnecessary testing" [9].

Passing on information within the Supply Chain and obligations of Downstream Users are administered in **Titles IV and V** [10, 11].

Title VI [12] is engaged in Evaluation of dossiers.

Authorizations are the subject of **Title VII** [13] and Restrictions are the topic for **Title VIII** [14].

Concerning the annexes it is important to know that **Annex IV and V** contain "Exemptions from the obligation to register" [15, 16].

Annex VI [17] lists information requirements referred to in Article 10. Article 10 is entitled "Information to be submitted for general registration purposes".

In Section 2 of **Annex VI** can be found the requirements concerning analytical information and analytical methods that have to be submitted to ECHA with Inquiry dossiers and also with dossiers for a full registration.

In **Annexes VII to X** [18a,b,c,d] there are lists with standard information requirements depending on the tonnage band that is relevant for the registration of a certain substance.

Annex XI [19] is important in cases where the standard testing regime described in **Annexes VII to X** [18a,b,c,d] seems not to be appropriate and therefore a registrant intends to deviate from the standard testing regime.

Annex XIV [20] contains a list of substances subject to authorization. This section was an empty one at the time REACH entered into force, but in the ongoing process when substances are identified for becoming subject to authorization this section will be amended regularly. This annex is a sort of living document.

1.3
Purpose and Scope of REACH

As usual in regulations in the first chapter of the REACH regulation [1] we find a statement concerning aim and scope.

Table of contents

TITLE 1	**General issues**	*Articles 1 to 4*
	Chapter 1	Aim, scope and application
	Chapter 2	Definitions and general provision
TITLE 2	**Registration of substances**	*Articles 5 to 24*
	Chapter 1	General obligation to register and information requirements
	Chapter 2	Substances regarded as being registered
	Chapter 3	Obligation to register and information requirements for certain types of isolated intermediates
	Chapter 4	Common provisions for all registrations
	Chapter 5	Transitional provisions applicable to phase-in substances and notified substances
TITLE III	**Data Sharing and Avoidance of unnecessary testing**	*Articles 25 to 30*
	Chapter 1	Objectives and general rules
	Chapter 2	Rules for non-phase-in substances and registrants of phase-in substances who have not preregistered
	Chapter 3	Rules for phase-in-substances
TITLE IV	**Information in the Supply Chain**	*Articles 31 to 36*
TITLE V	**Downstream Users**	*Articles 37 to 39*
TITLE VI	**Evaluation**	*Articles 40 to 54*
	Chapter 1	Dossier evaluation
	Chapter 2	Substance evaluation
	Chapter 3	Evaluation of intermediates
	Chapter 4	Common provisions
TITLE VII	**Authorization**	*Articles 55 to 66*
	Chapter 1	Authorization requirement
	Chapter 2	Granting of authorizations
	Chapter 3	Authorization in the supply chain
TITLE VIII	**Restrictions on the manufacturing, placing on the market and use of certain dangerous substances and preparations**	*Articles 67 to 73*
	Chapter 1	General issues
	Chapter 2	Restriction process
TITLE IX	**Fees and charges**	*Article 74*
TITLE X	**Agency**	*Articles 75 to 111*
TITLE XI	**Classification and Labeling Inventory**	*Articles 112 to 116*
TITLE XII	**Information**	*Articles 117 to 120*
TITLE XIII	**Competent authorities**	*Articles 121 to 124*
TITLE XIV	**Enforcement**	*Articles 125 to 127*
TITLE XV	**Transitional and final provisions**	*Articles 128 to 141*

Figure 1.2 Short overview on the table of contents of the REACH regulation including the list of Annexes.

List of Annexes

ANNEX I	General provisions for assessing substances and preparing Chemical Safety Reports
ANNEX II	Guide to the compilation of Safety Data Sheets
ANNEX III	Criteria for substances registered in quantities between 1 and 10 tonnes
ANNEX VI	Exemption from the obligation to register in accordance with Article 2 (7) (a)
ANNEX V	Exemption from the obligation to register in accordance with Article 2 (7) (b)
ANNEX VI	Information requirements referred to in Article 10
ANNEX VII	Standard information requirements for substances manufactured or imported in quantities of 1 tonne or more
ANNEX VIII	Standard information requirements for substances manufactured or imported in quantities of 10 tonnes or more
ANNEX IX	Standard information requirements for substances manufactured or imported in quantities of 100 tonnes or more
ANNEX X	Standard information requirements for substances manufactured or imported in quantities of 1000 tonnes or more
ANNEX XI	General rules for adaptation of the standard testing regime set out in ANNEXES VII to X
ANNEX XII	General provisions for Downstream users to assess substances and prepare Chemical Safety Reports
ANNEX XIII	Criteria for the identification of persistent, bioaccumulative and toxic substances, and very persistent and very bioaccumulative substances
ANNEX XIV	List of substances subject to Authorization
ANNEX XV	Dossiers
ANNEX XVI	Socioeconomic analysis
ANNEX XVII	Restrictions on the manufacture, placing on the market and use of certain dangerous substances, preparations and articles

Figure 1.2 *(Continued)*

In Article 1 (1) [21] is stated "The purpose of this Regulation is to ensure a high level of protection of human health and the environment, including the promotion of alternative methods for assessment of hazards of substances, as well as the free circulation of substances on the internal market while enhancing competitiveness and innovation."

This statement sounds pretty good, but it is difficult to judge whether it is possible to increase the level of protection of human health and the environment while also competitiveness and innovation shall be influenced in a positive way.

Representatives of the chemical industry normally have the perception that REACH is a great burden starting with purchasing of raw materials but not ending in selling a product to the customers.

In REACH Article 1 (2) [21] it is mentioned that the REACH regulation lays down provisions that shall apply to the whole lifecycle of a substance from manufacture, placing on the market until the use of the substance [21].

It is very important that "this Regulation is based on the principle that it is for manufacturers, importers and downstream users to ensure that they manufacture, place on the market or use such substances that do not adversely affect human health or the environment. Its provisions are underpinned by the precautionary principle" [21]. That means that the responsibility for ensuring a high level of protection of human health and environment lies within the chemical industry. A company in this sense is not only responsible to take care of the workers within the company but also responsible to take care of the protection of consumers and the environment. However, obligations and responsibility do not end at the moment a substance is sold to customers, because a registrant under REACH has in the registration dossier also to take care to cover all identified uses and corresponding exposure scenarios of his customers and the whole supply chain until the end of the lifecycle of the substance to be registered.

European Authorities, namely ECHA, carry over the responsibility concerning all risks that result for human beings and the environment from manufacturing and use of a certain substance to the registrant.

In practical terms this means every company has to ensure a manufacturing process by using state-of-the-art technologies to reduce risks for human and environment to a minimum. Furthermore, passing on information concerning safe handling to all customers and all members of the supply chain in an appropriate way is required.

1.4
Other Regulations and Directives that are Important in the Context of REACH

Already in the subtitle of REACH [1] there are cited some Directives and Regulations that were influenced by REACH [1].

Whereas Directive 1999/45/EC [22] is only amended by REACH [1], there are other directives and regulations that were repealed at the time REACH [1] entered into force. Repealed were Council Regulation (EEC) No 793/93 [23], Commission Regulation (EC) No 1488/94 [24], Council Directive 76/769/EEC [25], Commission Directive 91/155/EEC [26], Commission Directive 93/67/EEC [27], Commission Directive 93/105/EC [28] and as well Commission Directive 2000/21/EC [29].

Within the 131 short sections in the beginning of REACH [1] are further remarks concerning the influence of REACH on other regulations and directives and *vice versa*. In Section 9 [30] there is given the statement that by assessment of the operation of the four main legal instruments governing chemicals in the Community (Council Directive 67/548/EEC [2], Council Directive 76/769/EEC [25], Directive 1999/45/EC [22], Council Regulation (EEC) No 793/93 [23]) were identified a number of problems, resulting in disparities between the laws, regulations and administrative provisions in Member States. As the mentioned problems are history in so far that REACH applies in all Member States and existing national laws are ancillary to EU Law we will not go into details within this book. However, it may be that there are still some disparities in case of on-site inspections in chemical companies as the inspections are still accomplished by Member State competent authorities and not by ECHA.

There are other laws that remain unaffected by REACH [1] for example, the application of Directives on worker protection and the environment, especially

Directive 2004/37/EC [31] and there are also laws that do not affect REACH, for example, Council Directive 76/768/EEC [32]. The phase-out of testing on vertebrate animals for the purpose of protecting human health is restricted to the use of a substance in cosmetics [33]. When a substance is also used for other purposes than the use in cosmetics and these uses fall under the scope of REACH there will be no chance to outflank testing requirements demanded by REACH.

Within the 131 short sections in the beginning of REACH [1] we also find hints on substances that are excluded from the scope of REACH. We will focus on the exemptions from REACH in Section 3.9 of this book, therefore these cases are not discussed in this chapter.

Because of practical reasons it seems useful to mention some further legal documents that are of importance in the context of REACH in separate sections even when they are not mentioned in REACH [1] itself.

The selection cannot be complete, but the chosen ones have great influence in daily business. It is highly recommended that you are aware of the fact that these regulations exist and in case it should be necessary you will be in a position to consult the legal text concerning the details.

1.4.1
Fees and Charges Payable to the European Chemicals Agency

Article 74 (1) of REACH [1] states that fees that are required according to several Articles within this piece of legislation shall be specified in a Commission Regulation. In REACH [1] itself there are no concrete prices defined, but Article 74 (3) gives criteria to be considered in fixing prices for diverse types of services that ECHA provides: "The structure and amount of the fees . . . shall take account of the work required by this Regulation to be carried out by the Agency and the competent authority and shall be fixed at such a level as to ensure that the revenue derived from them when combined with other sources of the Agency's revenue pursuant to Article 96(1) is sufficient to cover the cost of the services delivered." It is also laid down that there should be reduced fees for SMEs. The structure and amount of fees, furthermore, shall take into account whether the registration is done jointly or as a single submission.

Circumstances under which a proportion of the fees will be transferred to the relevant Member State competent authority should be based on REACH Article 74 (4) and also be considered in the piece of legislation dealing with the fees order.

On 16 April 2008 Commission Regulation (EC) No 340/2008 dealing with fees and charges payable to the European Chemicals Agency entered into force. As we will see some details concerning fees payable for diverse types of registrations and some further services as for example, PPORD and Authorization in other chapters of this book whenever suitable, here no details shall be provided. However, this section can be closed with the remark that figures that were given within the fee order from 2008 were already amended for the first time in March 2013. The amendment was done in accordance with the statement of the former valid regulation in Section (17): "Fees and charges provided for under this Regulation should be adapted to take account of inflation and for that purpose the European Index of Consumer Prices published by Eurostat pursuant to Council Regulation (EC) No 2494/95 of 23 October 1995 concerning harmonized indices of consumer prices should be used."

Dated 20 March 2013 the former valid Regulation (EC) No 340/2008 was amended. Now the COMMISSION IMPLEMENTING REGULATION (EU) No 254/2013 [34] is valid.

1.4.2
Competition Law

REACH demands that companies that pre-registered the same substance and intend to register this substance should cooperate aiming to prepare a Joint submission. Information on the intrinsic properties of a substance to be registered jointly has to be shared. Data sharing is demanded especially in the case of studies that have been done by using vertebrate animals. It is clear that financial compensation within a consortium should take place – it shall be determined in a fair, transparent and non discriminatory way [35].

REACH does not give details on how companies can fulfill the single tasks, but you have to be aware of the fact that REACH is not a Competition-law-free zone.

That means *inter alia* that you are not allowed to abuse meetings concerning REACH matters for illegal actions such as speaking about prices of your products or any other information that has to be handled as confidential business information in accordance with the competition law. The above-mentioned examples may be obvious to everybody, but there may occur other situations that are not so clear. In case costs within a consortium shall be divided among the members based on the individual tonnage of a substance produced by every member, it is necessary to have an independent trustee, for example, a consultant company acting on behalf of the consortium dealing with confidential information that is not given to other members of the consortium.

In some cases, providing a Substance Information Profile for doing the Sameness-Check within a consortium can cause difficulties for example, in the case that from identified by-products or impurities can be traced back to the manufacturing process. On the other hand, you are obliged to ensure that Sameness is given to prove that one can assume that studies/tests done with the substance from company A is as similar to the substance from Company B that Company B does not have to repeat the studies/tests with the substance as manufactured within company B. To avoid trouble with the competition law, it is recommended to provide less detailed Substance Identification profiles. Very often it will be justifiable to agree on the main constituent and no impurity present that would be relevant for classification of the substance.

It could even be critical to think about preparing eSDS to be shared within the consortium. In general, it seems useful to prepare an eSDS jointly within a consortium as it is less work for each member and there will be a chance to agree on a harmonized template. If several companies within a consortium are willing to save money, because their customers are located only in a few EU countries and therefore the eSDS would be necessary only in a few languages, it will not be in accordance with the competition law to tell other members of the consortium in which countries your customers are located. It is highly recommended to provide either only an English version of the eSDS within a consortium to have a harmonized template and leave the necessary translations to each member or to provide the eSDS in all 23 EU languages.

As the Competition Law comprises several hundred pages and tasks under REACH are also very complex, it seems impossible to list all possible pitfalls within this chapter that may occur when you try to respect the Competition Law while fulfilling your obligations under REACH.

Whenever you are unsure whether there could arise any problem concerning competition law by doing your business for REACH, please check supporting Guidance documents [36], check the Competition Law in depth, ask a lawyer or employ a competent trustee.

1.4.3
GHS and CLP

The CLP regulation [37] incorporates the internationally agreed GHS criteria into Community Law. Within 79 short Sections in the beginning of this regulation one can find important remarks and references to other regulations that either have an influence on CLP or *vice versa*.

In Section 5 of the CLP regulation [38] we find a short note concerning the historical development of GHS: "With a view to facilitating worldwide trade while protecting human health and the environment, harmonized criteria for classification and labeling have been carefully developed over a period of 12 years within the United Nations (UN) structure, resulting in the Globally Harmonized System of Classification and Labeling of Chemicals (hereinafter referred to as 'the GHS')" [38]. In the next Section is mentioned: "This Regulation follows various declarations whereby the Community confirmed its intention to contribute to the global harmonization of criteria for classification and labeling, not only at UN level, but also through the incorporation of the internationally agreed GHS criteria into Community law" [39].

The first crossreference to REACH is made in Section 12: "The terms and definitions used in this Regulation should be consistent with those set out in Regulation (EC) No 1907/2006 of the European Parliament and of the Council of 18 December 2006 concerning the Registration, Evaluation, Authorization and Restriction of Chemicals (REACH), with those set out in the rules governing transport and with the definitions specified at UN level in the GHS, in order to ensure maximum consistency in the application of chemicals legislation within the Community in the context of global trade. The hazard classes specified in the GHS should be set out in this Regulation for the same reason" [40].

In our daily business concerning REACH matters we have at least two situations where REACH is touched by GHS/CLP and *vice versa*.

First, we are obliged to fill in the Chapter 2 of a registration dossier information concerning classification and labeling in accordance with GHS and the obligation to submit a CLP notification can be fulfilled by submitting this information included in the registration dossier. For registrations done until 30 November 2010 by an Only Representative of a Non-EU manufacturer the inclusion of the classification and labeling in accordance with GHS meant that EU customers marketing the registered substance later did not have an obligation to do an CLP notification on their own. If an Only Representative will register at a later stage (registration deadline e.g., in 2013) EU customers purchasing from the corresponding Non-EU manufacturer and intending to market a certain substance had to do their own CLP notification until 3 January 2011.

The second situation were REACH meets CLP in the daily business of the chemical industry, is the preparation of Safety Data Sheets. There are several amendments in Safety Data Sheets compared to former days that have to be done to fulfill obligations concerning CLP. REACH [1] describes several demands concerning preparation and content of Safety Data Sheets in Article 31. Normally you can find the uses of a substance as registered in the Safety Data Sheet and in the case of standard registrations there will be also an Annex to the so-called "extended Safety Data Sheet" (eSDS) including Exposure scenarios.

1.4.4
Other Regulations Containing the Wording REACH

In almost every country of the world there are existing laws commanding conditions for manufacturing and use of chemicals. In many states manufacturer, of chemicals also have to inform national authorities concerning products that are manufactured or used at each site of a company. Data requirements may vary from country to country, but all these regulations have in common that protection of human health and the environment is important in the perception of the public. However, even when there is included the wording "REACH" in the name of such laws, as it is for example, in China REACH or Korea REACH, it has nothing to do with the European REACH regulation [1], although the EU REACH may have been a sort of benchmark example in the development of some national laws concerning chemicals.

So far, contrary to GHS, there have been no efforts made to implement a unique "REACH" legislation for the whole world.

References

1 Regulation (EC) No 1907/2006 of the European Parliament and of the Council of 18 December 2006, concerning the Registration, Evaluation, Authorisation and Restriction of Chemicals (REACH), establishing a European Chemicals Agency, amending Directive 1999/45/EC and repealing Council Regulation (EEC) No 793/93 and Commission Regulation (EC) No 1488/94 as well as Council Directive 76/769/EEC and Commission Directives

91/155/EEC, 93/67/EEC, 93/105/EC and 2000/21/EC.
2 Council Directive of 27 June 1967 on the approximation of laws, regulations and administrative provisions relating to the classification, packaging and labelling of dangerous substances (67/548/EEC).
3 EINECS = European Inventory of Existing Commercial Chemical Substances, is available within the ESIS database under http://esis.jrc.ec.europa.eu/
4 See Title of [1] and within this regulation also Section 15 of the determinations made in the beginning of this regulation.
5 See Article 5 of the REACH regulation [1].
6 See Article 1 and Article 2 of the REACH regulation [1].
7 See Article 3 and Article 4 of the REACH regulation [1].
8 See Articles 5 to 24 of the REACH regulation [1].
9 See Articles 25 to 30 of the REACH regulation [1].
10 See Articles 31 to 36 of the REACH regulation [1].
11 See Articles 37 to 39 of the REACH regulation [1].
12 See Articles 40 to 54 of the REACH regulation [1].
13 See Articles 55 to 66 of the REACH regulation [1].
14 See Articles 67 to 73 of the REACH regulation [1].
15 See Annex IV "Exemptions from the obligation to register in accordance with Article 2 (7) (a)" of the REACH regulation [1].
16 See Annex V "Exemptions from the obligation to register in accordance with Article 2 (7) (b)" of the REACH regulation [1].
17 See Annex VI "Information requirements referred to in Article 10" of the REACH regulation [1].
18 (a) See Annex VII "Standard information requirements for substances manufactured or imported in quantities of 1 tonne or more" of the REACH regulation [1]; (b) See Annex VIII "Standard information requirements for substances manufactured or imported in quantities of 10 tonnes or more" of the REACH regulation [1]; (c) See Annex IX "Standard information requirements for substances manufactured or imported in quantities of 100 tonnes or more" of the REACH regulation [1]; (d) See Annex X "Standard information requirements for substances manufactured or imported in quantities of 1000 tonnes or more" of the REACH regulation [1].
19 See Annex XI "General rules for adaptation of the standard testing regime set out in Annexes VII to X" of the REACH regulation [1].
20 See Annex XIV "List of substances subject to authorisation" of the REACH regulation [1].
21 See Article 1 of the REACH regulation [1].
22 Directive 1999/45/of the European Parliament and of the Council of 31 May 1999 concerning the approximation of the laws, regulations and administrative provisions of the Member States relating to the classification, packaging and labelling of dangerous preparations. Directive as last amended by Commission Directive 2006/8/EC.
23 Council Regulation (EEC) No 793/93 of 23 March 1993 on the evaluation and control of the risks of existing substances. Regulation as amended by Regulation (EC) No 1882/2003 of the European Parliament and of the Council.
24 Commission Regulation (EC) No 1488/94 – laying down the principles for the assessment of risks to man and the environment of existing substances in accordance with Council Regulation (EEC) No793/93.
25 Council Directive 76/769/EEC of 27 July 1976 on the approximation of the laws, regulations and administrative provisions of the Member States relating to restrictions on the marketing and use of certain dangerous substances and preparations. Directive as last amended by Directive 2005/90/EC of the European Parliament and of the Council.
26 Commission Directive 91/155/EEC of 5 March 1991 defining and laying down the detailed arrangements for the system of specific information relating to dangerous preparations in implementation of Article

10 of Directive 88/379/EEC – as amended by Commission Directive 93/112/EC of 10 December 1993.
27. Commission Directive 93/67/EEC of 20 July 1993 laying down the principles for assessment of risks to man and the environment of substances notified in accordance with Council Directive 67/548/EEC.
28. Commission Directive 93/105/EC of 25 November 1993 laying down Annex VII D, containing information required for the technical dossier referred to in Article 12 of the seventh amendment of Council Directive 67/548/EEC.
29. Commission Directive 2000/21/EC of 25 April 2000 concerning the list of Community legislation referred to in the fifth indent of Article 13(1) of Council Directive 67/548/EEC.
30. See Section 9 in the beginning of REACH [1].
31. Directive 2004/37/EC of the European Parliament and of the Council of 29 April 2004 on the protection of workers from the risks related to exposure to carcinogens or mutagens at work.
32. Council Directive 76/768/EEC of 27 July 1976 on the approximation of the laws of the Member States relating to cosmetic products. Directive as last amended by Commission Directive 2005/80/EC.
33. See Section 13 in the beginning of REACH [1].
34. COMMISSION IMPLEMENTING REGULATION (EU) No 254/2013 of 20 March 2013 amending Regulation (EC) No 340/2008 on the fees and charges payable to the European Chemicals Agency pursuant to Regulation (EC) No 1907/2006 of the European Parliament and of the Council on the Registration, Evaluation, Authorisation and Restriction of Chemicals (REACH).
35. See Article 27 (3) and Article 30 (1) of REACH [1].
36. ECHA, Guidance on data sharing, Version 2.0, dated April 2012.
37. Regulation (EC) No 1272/2008 of the European Parliament and of the council of 16 December 2008 on classification, labeling and packaging of substances and mixtures, amending and repealing Directive 67/548/EEC and 1999/45/EC, and amending Regulation (EC) No 1907/2006.
38. See Section 5 in the beginning of the CLP regulation [37].
39. See Section 6 in the beginning of the CLP regulation [37].
40. See Section 12 in the beginning of the CLP regulation [37].

2
Roles under REACH

A chemical company can have more than one role under REACH. The number of roles an enterprise can have depends on the location of the legal entity manufacturing a certain substance (within the EU or outside of the EU) and also on the sort of substance that is dealt with: raw material that is purchased from a supplier or a substance that is manufactured by the company itself.

In general, there are only cases to be considered in which substances are manufactured in the European Union or imported to the European Union in amounts of 1 t/a or more per manufacturer or importer.

Case by case there can arise different obligations from the special situation. It is also possible that there is more than one option to achieve REACH compliance. Every company has to check the situation for every substance that is handled within this company and sometimes a situation will occur in which a discussion between different parties within the supply chain is necessary to ensure that all obligations are fulfilled in accordance with the REACH regulation.

For enterprises that have several legal entities (sites in the EU and sites outside of the EU) the question can occur whether a single substance should be produced at a European site or somewhere outside of the EU. In general, such business decisions will not be based only on REACH matters, but REACH can have a certain influence among other details.

2.1
Manufacturer within the EU

According to Article 3 (9) of the REACH regulation [1] "Manufacturer: means any natural or legal person established within the Community who manufactures a substance within the Community;" and Article 3 (8) defines: "Manufacturing: means production or extraction of substances in the natural state" [2].

Every manufacturer of a chemical substance who is located in the European Union has the obligation to register this substance in accordance with REACH when that substance is manufactured in quantities of 1 t/a or more by this manufacturer.

There are only a few exemptions from the general obligation to register a chemical substance. In cases where a chemical substance is not isolated, it is covered by another legislation or can be considered as waste an EU manufacturer may have no registration obligation.

Actually, there are 28 Member states within the European Union in which REACH directly applies. As already mentioned in Chapter 1 the European Economic Area (EEA) comprises the 28 Member states of the European Union plus Norway, Iceland and Liechtenstein. Therefore, to be exact there are 31 states of EEA that adopt the REACH regulation within their territory. Within this book "EU Manufacturers" will also include Manufacturers located in Norway, Iceland and Liechtenstein, but does not include any member located in a state that does not belong to the EEA.

All manufacturers that are located outside of the European Union (and are not members of the EEA) are Non-EU manufacturers. Switzerland is located on the European continent but it is not a member of the EEA, therefore manufacturers producing in this country are Non-EU manufacturers in terms of REACH.

2.2
Non-EU Manufacturer, Importer and Only Representative

All manufacturers that produce outside of the European Union do not have any obligation to register any substance under REACH.

However, when they sell their products to customers within the EU somebody has to take care that at least the amounts of a substance that are imported into the EU are (pre-)registered. In terms of REACH Article 3 (10) defines: "Import: means the physical introduction into the customs territory of the Community" [3].

There are in general two options to fulfill the registration obligation:

The customer within the EU has the role of the importer under REACH. "Importer: means any natural or legal person established within the Community who is responsible for import" as defined by REACH Article 3 (11) [4].

The importer is obliged to take care that all obligations concerning REACH are fulfilled. The only possibility to deviate from this general rule is given in case the Non-EU manufacturer takes over the registration obligations as a Non-EU manufacturer is allowed to do registrations via an Only Representative (OR) within the EU [5]. The Only Representative acts formally as the importer under REACH and fulfills the registration obligations. The customers of this Non-EU manufacturer can benefit in so far that they are Downstream Users [6] and do not have any registration obligation for the amounts of this substance that they purchase from this single Non-EU manufacturer. It is foreseen that a Non-EU manufacturer that will register via Only Representative informs his customers concerning identity and address of the Only Representative [6].

The Only Representative shall comply with all other obligations of importers under the REACH regulation [7]. He shall keep available and up-to-date informa-

2.2 Non-EU Manufacturer, Importer and Only Representative

tion on quantities imported and customers sold to, as well as information on the supply of the latest update of the safety data sheet [7]. Within a globally acting company having separate independent legal entities in countries within the EU and also outside of the EU the role of the Only Representative can be fulfilled by a legal entity located within the EU and all necessary data may be available for the Only Representative within globally available company-based IT-systems, therefore it will be easier within this company group to deal with the obligations an Only Representative has than in the case of a Non-EU company having an Only Representative within the EU that acts on behalf of this manufacturer but is working independently of this company (e.g., a consultant acting as an Only Representative). A consultant acting as an Only Representative on behalf of several companies will have a higher expenditure concerning administrative work and data exchange with the Non-EU customer than an Only Representative within a company group that has an ideal IT-platform at its disposal.

It is absolutely important that companies located within the EU that purchase raw materials from Non-EU manufacturers find out for every single source whether they have an obligation to register or not.

It is recommended to have in every company an overview comprising at least all raw materials that are purchased from Non-EU sources. Please find an example of an appropriate list in Figure 2.1.

In case you have to ask your suppliers concerning the REACH status of a product purchased by your company you can either send a short email to your supplier or you can send a letter and a questionnaire to your supplier. In any case it is recommended to give your supplier as much information that he will be in the position to answer your questions within a few minutes. This means that you should clearly indicate for which substance you request information (substance

Substance name	CAS	EINECS	My company has pre-registered (tonnage band)	Supplier	Amount purchased per year by this supplier	Questionnaire concerning REACH status sent to supplier on	Supplier has pre-registered?	Supplier intends to register?	My company Will be Downstream User or Importer?	TO DO for my company
Substance S			[Yes/no] (...t/a)	1	X t/a	[date]	[Yes/no]	[Yes/no]	[DU/I]	
				2	Y t/a					
				3	Z t/a					
				4						
				5						

Figure 2.1 Overview Raw Materials, suppliers and REACH status.

name, CAS number, EC number). Very often personnel in REACH departments receive requests in which the customers ask them to list all products that are supplied to this customer and to give information on the REACH status of all these products. This will either result in no response, because personnel in REACH Departments do not have directly access to the customer and product relationship, or it will be extremely time consuming and maybe you will have to wait for a long time until you receive an answer. Please, do not list substances that you bought 20 years ago but that your company does not buy any more. A useful example of correspondence to your supplier can be found in Figure 2.2.

Example Fine Chemicals Company
Fine Chemicals Street 12
P.O. Box
67845 Example City
Finland

Fast Solutions
Mr James Miller
Sales Manager
17 Finsbury Square
USA

[Date]

Dear Mr Miller/Dear supplier,

Request for REACH status of substance(s) purchased from your company

As a company located within the European Union we have to take care of all registration obligations under REACh concerning substances manufactured at our site and also for all substances imported from Non-European suppliers, if they do not take care of the REACh registration obligations via an Only Representative within the EU.
Therefore, we would like to ask you whether your company has pre-registered/intends to register Substance S, CAS [ ], EINECS [ ] under REACH via an Only Representative within the EU.
Could you please be so kind as to fill in the enclosed questionnaire and send it back at your earliest convenience.

Best regards,

Enc.: Questionnaire concerning REACH status of Substance S

Figure 2.2 Requests to your supplier concerning REACH status of products purchased by this supplier – letter to supplier and questionnaire.

Figure 2.2 *(Continued)*

Very often, EU customers expect that their Non-EU suppliers take care of all registration obligations as EU suppliers have to. When a Non-EU manufacturer is not willing to provide his customers with this sort of service he can lose customers and large parts of his business. To keep in line with the real market conditions a Non-EU manufacturer who intends to act globally has to be aware of the REACH regulation and the influence on his business although there is no direct obligation to register from the legal point of view. If a Non-EU Manufacturer who has customers within the EU and also customers outside of the EU intends to register under REACH he has only to consider the amounts of the substance that are brought into the EU. This is a big difference in comparison to an EU manufacturer who has the obligation to register the complete amount of a substance that is manufactured at his site even when the customer is outside the EU.

When the Non-EU manufacturer does a registration via Only Representative (OR) for a substance in a lower tonnage band than the highest one, it is highly recommended to do a Substance Volume Tracking. This can either be implemented by appropriate IT-systems or with a suitable table. A spreadsheet appropriate for doing Substance Volume Tracking can be found in Figure 2.3.

When the tonnage band has to be increased the registrant is obliged to update his registration without undue delay.

2.3
Downstream User

A Downstream User is any natural or legal person established within the Community, other than the manufacturer or the importer, who uses a substance, either on its own or in a preparation, in the course of his industrial or professional activities [8]. A distributor or a consumer is not a downstream user [9]. A re-importer shall be regarded as a downstream user [8].

It is clear that every company purchasing a chemical compound from a European Manufacturer will be a Downstream User for the amounts of this substance that are purchased from European Manufacturers.

When an Importer cares for the REACH registration obligations and afterwards sells a certain substance to his customers, all his customers will be Downstream Users.

A Downstream User in the meaning of REACH can be a company located within the EU even when this company purchases a certain substance from a Non-EU manufacturer. If the Non-EU manufacturer is willing to register a substance via an Only Representative (having its place of business within the EU) the customer(s) of this Non-EU manufacturer will be a Downstream User(s) without any doubt. Even when the EU customer comes and fetches the product from the Non-EU manufacturer's site and therefore is doing the physical import on their own railroad trucks the EU customer will be a Downstream User in the meaning of REACH. However, in this special case the customer will have to take care of the duties concerning clearance in accordance with the terms of delivery.

Substance	CAS	EINECS	Tonnage band for preregistration	CMR or R50/53?	Amount manufactured/ imported in 2004	2005	2006	2007	2008	2009	2010	2011	Average tonnage	Registration deadline	TO DO

Remark: If a substance has been manufactured/imported for at least three consecutive calendar years, the estimated tonnage band in the following calendar year can be calculated based on the average of the last three calendar years [8] [10] see Article 3 (30) of the REACH regulation [1]. Registration deadline is given by the highest tonnage band per calendar year after June 2007. Data requirements are based on the relevant average tonnage band of the last three calendar years before the date of registration.

Figure 2.3 Substance Volume Tracking.
For Manufacturers/Importers within the EU total amount produced/imported per calendar year have to be considered
For Non-EU Manufacturers that register via Only Representative within the EU only amounts that are sold to European customers have to be considered

The Only Representative has to be informed by the Non-EU manufacturer concerning the amounts that are delivered to European customers but the physical import can go directly from the Non-EU manufacturer to the customer(s) within the EU.

"Under REACH, downstream users must not place on the market or use any substances which are not registered in accordance with REACH" [11] Guidance for downstream users.

This means a Downstream User handles only substances that were

- produced/imported by the supplier in amounts below 1 t/a;
- have already been registered by the manufacturer or importer;
- have been (late) pre-registered by the manufacturer or importer and there is still time until the registration deadline;
- have been registered or (late) pre-registered by an Only Representative of a Non-EU manufacturer.

The first obligation of a Downstream User to be therefore will be to ensure that his role within the supply chain really is that of a Downstream User.

Therefore, you should find out whether your supplier is aware of REACH and has taken all necessary actions. To fulfill this task you may use the spreadsheet in Figure 2.1 and the letter to suppliers as shown in Figure 2.2.

Although a company that found out that they are Downstream User for a certain substance does not have any registration obligations concerning REACH, there are several other obligations to be considered.

When a supplier intends to register a substance as a transported isolated intermediate, he will need a Confirmation from all customers that they use this certain substance for the synthesis of another substance and always abide strictly controlled conditions as listed in Article 18 (4) [12]. The Downstream User will have to check the situation at his own site and when he is willing to support his supplier with the mentioned confirmation he will be responsible for abiding by the conditions as demanded in Article 18 (4). The supplier will not have control at the customer's site as the registrant has only the obligation to ask the customers for the confirmation and to abide the strictly controlled conditions at the manufacturing site if he wants to benefit from the cheaper type of registration as a transported isolated intermediate. But it is possible that there will be on-site inspections at the customer's site, during which the national authorities will control whether the Downstream User's handling is in compliance with the REACH regulation.

If a Downstream User has a certain use that shall be considered in the registration dossier, the Downstream User may inform his supplier. As a minimum the brief general description of use shall be passed on in writing (on paper or electronically) to the manufacturer, importer, downstream user or distributor who supplies him with the aim of making this an identified use [13].

Sometimes, the registrant to be will then need some further support from the Downstream User concerning Exposure scenarios or use and exposure category to be considered in the Chemical Safety Assessment of his supplier [13].

Distributors shall pass on such information to the next actor or distributor up the supply chain [13]. The request of the Downstream User shall be made in due time. In due time means that for phase-in substances the registrant to be should have this information at least twelve months prior to his registration deadline [14].

For registered substances it is foreseen that the supplier has to consider requests from Downstream Users before he next supplies the substance, provided the request was made at least one month before the supply, or within one month after the request, whichever is the later [14].

When a registrant is unable to include the use of the Downstream User as an identified use for reasons of protection of human health or the environment, he shall provide ECHA and the downstream user with the reason(s) for that decision in writing without delay and shall not supply downstream users with the substance without including these reasons in the SDS [14].

For any use outside the conditions described in an exposure scenario or if appropriate a use and exposure category communicated to him in a SDS or for any use his supplier advises against, a downstream user in general shall prepare a chemical safety report [15], although there are some exemptions [15].

When a Downstream User has the obligation to prepare a Chemical Safety Report, he has to do this within 12 months after having received a registration number communicated to them by their suppliers in a SDS [16]. ECHA has to be informed at the latest six months after the Downstream user received a registration number communicated to them by their suppliers in a SDS [17]. The content of the information that ECHA expects is determined. Where it is considered necessary by the Downstream User to complete his chemical safety assessment, a proposal for additional testing on vertebrate animals can be included [18]. A downstream user is also obliged to study carefully SDS or any information on risk-management measures supplied to him with the aim to identify, apply and where suitable, recommend, appropriate measures to adequately control identified risks [19]. In Figure 2.4 you will find an overview on the most important obligations and tasks of a downstream user.

2.4
Trader within EU versus Non-EU Trader and Distributor

A trader is an enterprise that does not manufacture a certain substance itself, but purchases this substance from a manufacturer and afterwards sells the whole amount or maybe repacks the substance in smaller packages and afterwards sells them to several customers.

REACH differentiates between European traders and Non-EU traders.

A Non-EU trader is not allowed to register at all. Contrary to a Non-EU manufacturer he is not allowed to register via an Only Representative in the EU. This means that customers purchasing raw materials from a Non-EU trader may have to face the fact that they have to register in the role of the importers.

Obligations & Tasks	Supporting documents
1. Find out and define your role under REACH concerning every single substance purchased by your company from any supplier Further actions could be: • looking for another supplier if the former one will not register • change of your role from Downstream user to importer (including registration obligations)	• see list in Figure 2.1 • letter to supplier in Figure 2.2 • [9]
2. Cooperation with supplier/registrant concerning preparation of the registration dossier • Find out how every single substance that you purchase is used in your company (Intermediate, enduse, solvent, catalyst) and find out details concerning Exposure scenarios • Communicate your uses to your supplier in due time to make them "identified uses" • Offer studies/tests for that your company is Data Holder to your suppliers in advance of the preparation of the registration dossier	
3. Answer to requests from your supplier/the registrant • If your supplier requests an Art. 18 (4) confirmation, please cooperate with him whenever possible as registration as a transported isolated intermediate is cheaper than a full registration and your company as a customer will also benefit from cheaper costs	• [10]

4. Communication with your supplier after registration of a certain substance
• On receipt of a new SDS or an eSDS
  • Check whether the identified uses of your company are covered
    - When a particular use is not covered, contact your supplier and ask for inclusion in the registration dossier of the registrant
    - In case your use is advised against, please inform ECHA and prepare your own Chemical Safety Assessment in due time
  • apply Risk-Management Measures as foreseen in the SDS/eSDS of your supplier
	• [9]

5. Communication with your customers
• If you sell a substance that you purchased from a supplier to your customers (down the supply chain)
  - Provide your customers with an appropriate SDS/eSDS in due time
  - Ensure that the registration number of the registrant has to be shortened to the format 01-10 digits-2 digits (the last four digits that can be used to trace back to the registrant shall be omitted to be in compliance with the Competition Law)
	• [11]

Figure 2.4 CHECK-LIST for downstream users.

For a trader located within the EU there are some more options to cope with REACH registration obligations than those for his Non-EU colleagues.

A European trader can purchase a substance from a European manufacturer. In this case the trader/his customers can be regarded as Downstream Users without any registration obligations.

The European trader can also buy from a Non-EU manufacturer that registers via an Only Representative in the EU. In this case, the EU trader/his customers can also be regarded as Downstream Users without registration obligations.

When a European trader buys from a Non-EU manufacturer or a Non-EU trader, the EU trader can also act as an importer in matters of REACH. In this case he is allowed and obliged to take care of the registration obligations.

A trader may buy a substance in a tank container and afterwards have the idea to refill/repack the substance in small bottles. The refilling/repacking in the meaning of REACH is an identified use [20] and therefore the mentioned trader in this case maybe a Downstream User. However, he will not be a Distributor in the meaning of REACH, because it is defined that a "Distributor: means any natural or legal person established within the Community, including a retailer, who only stores and places on the market a substance, on its own or in a preparation, for third parties;" [21]. This means a trader who buys five barrels of a substance A, stores them for a certain time and afterwards sells them to his customers without opening the barrels, can be regarded as a Distributor. A Distributor therefore will not be a Downstream User in the meaning of REACH [22].

2.5
Examples and Exercises

1. Company A is a manufacturer located in the Netherlands, company B is a manufacturer located in Switzerland. Both produce a substance S in amounts of 100 t/a.

   Who has registration obligations?

2. Company A located in Germany manufactures as Company B in Switzerland 10 t/a of a substance S.

   a) Both companies sell the full amount of S to a customer C located in the USA. Who has registration obligations?

   b) Both companies sell the full amount to a customer C located in Italy. Who has registration obligations?

   c) Both companies sell the substance S to 12 customers located in diverse EU countries. Each of this customers purchases less than 1 t/a of the substance S. Who has registration obligations?

3. It is November 2011. Company A located in Belgium wants to import 2 t of a substance S from China for the first time. This substance is a non-phase-in substance. The substance will be used in the end of January 2012. The Chinese Manufacturer did not cooperate with an Only Representative within the EU and does not intend to do any activities concerning REACH. How can company A achieve REACH compliance?

References

1 Regulation (EC) No 1907/2006 of the European Parliament and of the Council of 18 December 2006, concerning the Registration, Evaluation, Authorisation and Restriction of Chemicals (REACH), establishing a European Chemicals Agency, amending Directive 1999/45/EC and repealing Council Regulation (EEC) No 793/93 and Commission Regulation (EC) No 1488/94 as well as Council Directive 76/769/EEC and Commission Directives 91/155/EEC, 93/67/EEC, 93/105/EC and 2000/21/EC.
2 See Article 3 (8) of the REACH regulation [1].
3 See Article 3 (10) of the REACH regulation [1].
4 See Article 3 (11) of the REACH regulation [1].
5 See Article 8 (1) of the REACH regulation [1].
6 See Article 8 (3) of the REACH regulation [1].
7 See Article 8 (2) of the REACH regulation [1].
8 See Article 3 (13) of the REACH regulation [1].
9 See Article 3 (14) of the REACH regulation [1].
10 See Article 3 (30) of the REACH regulation [1].
11 ECHA, Guidance for downstream users, January 2008, page 13.
12 See Article 18 (4) of the REACH regulation [1].
13 See Article 37 (2) of the REACH regulation [1].
14 See Article 37 (3) of the REACH regulation [1].
15 See Article 37 (4) of the REACH regulation [1].
16 See Article 39 (1) of the REACH regulation [1].
17 See Article 39 (2) of the REACH regulation [1].
18 See Article 38 (2) of the REACH regulation [1].
19 See Article 37 (5) of the REACH regulation [1].
20 See Article 3 (24) of the REACH regulation [1].

3
What Sort of Substances have to be Considered under REACH

3.1
Substance, Mixture and Article under REACH

Substance, mixture and article are defined in Article 3 of the REACH regulation [1] and it is essential to have a clear understanding of these definitions as under REACH only substances can be registered.

REACH Article 3 (1) states: "**Substance**: means a chemical element and its compounds in the natural state or obtained by any manufacturing process, including any additive necessary to preserve its stability and any impurity deriving from the process used, but excluding any solvent which may be separated without affecting the stability of the substance or changing its composition;" [2].

Mercury is a substance. Dichloromethane, chlorine, sodium hydroxide and hydrogen chloride are also substances.

REACH Article 3 (2) defines: "**Preparation**: means a mixture or solution composed of two or more substances;" [3]

An example of such a preparation are all substances dissolved in a solvent, for example, sodium hydroxide in water. A mixture of ethanol and methanol is also a preparation within the meaning of REACH. In general, a preparation is made by mixing of two or more substances that do not react with each other. Normally these substances can be separated by using appropriate techniques for example, distillation.

In REACH Article 3 (3) is mentioned: "**Article**: means an object which during production is given a special shape, surface or design which determines its function to a greater degree than does its chemical composition;" [4]

Examples of an article within the meaning of REACH can be found in great numbers in our surroundings – maybe you are sitting on a chair in this moment. A chair is an article as it does not matter whether it is made from wood, metal or plastic. Also a vacuum cleaner and knife, fork and spoon are articles.

Under REACH it is important to know the composition of a substance and also how the substance will be used during its lifecycle.

As we saw in the beginning of this chapter a substance could have a purity of 100%, but this is not convincing. In reality, very often the situation will occur that based on the purity of the raw material used and the type of reaction the resulting product will show a purity of less than 100%. Depending on the composition REACH differentiates between mono-constituent substances, multi-constituent substances and Substances of Unknown or Variable composition, Complex reaction products or Biological materials (UVCB).

Independently of this sort of classification there can be different types of use for every substance that have a direct influence on the type of registration that will be needed.

3.2
Different Compositions

Based on the knowledge about their composition substances can be divided into two main groups [5]:

"Well-defined substances" and Substances of Unknown or Variable composition, Complex reaction products or Biological materials ("UVCB substances").

"Well-defined substances" are substances with a defined qualitative and quantitative composition [5]. There are two subtypes of well-defined substances: mono-constituent substances and multi-constituent substances. In general it is possible to characterize substances of these two types by well-known analytical methods.

In some special cases additional parameters such as crystallomorphology can be important, for example, for differentiation between graphite and diamond [5].

3.2.1
Mono-constituent Substance

A mono-constituent substance is defined as a substance with just one main constituent. The content of this main constituent has to be ≥80% (w/w) [5]. The name of this substance is given after the main constituent.

When there is no impurity relevant for classification and labeling the substance sameness for this substance produced by two or more manufacturers is given by the fact that all of these manufacturers sell a substance with this certain main constituent.

3.2.2
Multi-constituent Substance

A multi-constituent substance is a substance, defined by its quantitative composition, in which more than one main constituent is present in a concentration >10% (w/w) and <80% (w/w) [5].

Contrary to a mixture that is made by mixing together several substances that do not react with each other a multi-constituent substance is the product of a manufacturing process.

Therefore, a multi-constituent substance is named as "Reaction mass of (names of the main constituents)" [5].

Very often a multi-constituent substance is obtained from reaction types such as rearrangements, eliminations or olefin metathesis.

3.2.3
Substances of Unknown or Variable Composition, Complex Reaction Products or Biological Materials

As Substances of Unknown or Variable composition, Complex reaction products or Biological materials (UVCB) cannot easily be identified by their composition, in general the source of the substance and the process used are of high importance. The terms "main constituents" and "impurities" should not be regarded as relevant for UVCB substances [5].

If a mixture of fatty acids from natural sources is transformed into acid chlorides, the composition of the product will depend on the composition of the fatty acid. The composition of the naturally occurring fatty acid for example, palm oil will have a variable composition based on place and year of harvest. Therefore, the name of the product could be "chlorinated fatty acids C_x–C_y" [5].

3.3
Different Types of Use

We saw that for identification and naming of a substance its composition is of high importance, but the use of the substance has a great influence on the type of registration that has to be considered.

3.3.1
Substance with End Use

A substance that is not consumed in a process intended to transform this substance into another substance, is a substance that cannot be described as an intermediate. It is a substance with end use. In this category of substances fall all solvents, catalysts and in general substances that are used by consumers. As there is a widespread use and risk for human health and the environment cannot be ignored, in such cases a full registration is required. For tonnage bands of 10 t/a and more manufacturers and importers are obliged to also provide a Chemical Safety Report included in the registration dossier.

The same procedure is necessary for monomers as the corresponding polymers are exempted from registration obligations.

Whereas for substances with end use a full registration is in any case obligatory, intermediates can be exempted from registration obligations or at least benefit from reduced data requirements provided that certain conditions are fulfilled.

3.3.2
Intermediate

In REACH Article 3 (15) an Intermediate is defined: this means a substance that is manufactured for and consumed in or used for chemical processing in order to be transformed into another substance (hereinafter referred to as "synthesis") [6]. Depending on the type of storage and depending on whether a substance is transported after the manufacturing process itself intermediates can be divided into three subcategories. Being classified into one of this subcategories has a direct influence on the registration obligations and if so the type of registration that is required.

3.3.2.1 Non-isolated Intermediate
In accordance with REACH Article 3 (15) (a) "non isolated intermediate: means an intermediate that during synthesis is not intentionally removed (except for sampling) from the equipment in which the synthesis takes place. Such equipment includes the reaction vessel, its ancillary equipment, and any equipment through which the substance(s) pass(es) during a continuous flow or batch process as well as the pipework for transfer from one vessel to another for the purpose of the next reaction step, but it excludes tanks or other vessels in which the substance(s) are stored after the manufacture;" [6].

By doing a multi-step synthesis either in one vessel or a cascade of vessels all non-isolated intermediates do not have to be registered – non-isolated intermediates are exempted from any registration obligation under REACH.

3.3.2.2 On-site Isolated Intermediate
REACH Article 3 (15) (b) states: "on-site isolated intermediate: means an intermediate not meeting the criteria of a non-isolated intermediate and where the manufacture of the intermediate and the synthesis of (an)other substance(s) from that intermediate take place on the same site, operated by one or more legal entities;" [6].

This means when a company isolates an intermediate and this intermediate is stored at this site and afterwards used at the same site we have an on-site isolated intermediate. In contrary to the non-isolated intermediate a Manufacturer located within the EU has to register such a substance, although he can benefit from reduced data requirements compared to a full registration. This is valid when the registrant can provide proof that he abides the strictly controlled conditions as stated in REACH Article 17 (3) [7].

If a Manufacturer located outside of the EU stores on-site isolated intermediates before using them for transformation into another substance there are no registration obligations at all.

This should be considered by decision makers in enterprises having sites within the EU and also outside the EU. By doing such steps outside the EU in some cases the intermediate steps can be hidden. That could be useful in case there would be confidential business information made public by the registration of several intermediates of a multi-step reaction, which could be traced back to the synthesis route.

On the other hand, when there are more dedicated facilities available within your European sites the registration of an on-site isolated intermediate is not too expensive provided your company is in the position to fulfill all requirements concerning strictly controlled conditions.

3.3.2.3 Transported Isolated Intermediate

Concerning a transported isolated intermediate REACH Article 3 (15) has only a short definition. "Transported isolated intermediate: means an intermediate not meeting the criteria of a non-isolated intermediate and transported between or supplied to other sites;" [6].

The definition itself may be short, but the topic is at least so complex that ECHA published the Guidance on intermediates [8] and although there is a second version available there are still many open questions.

It is clear that a manufacturer and also the user of a transported isolated intermediate have to abide the strictly controlled conditions as given in REACH Article 18(4) [9] to be in a position to benefit from reduced data requirements for their registration. If it is not possible to fulfill the conditions as stated in REACH Article 18(4) [9] it is clear that the registrant has to make efforts to achieve a full registration. Details will be discussed in Section 5.2 of this book.

3.4 Phase-In Substances

The former orphans in the shape of already-known substances attained a special importance at the moment when REACH entered into force.

Special attention now lies on the existing commercial chemical substances that are listed in the EINECS [10] and had been on the market before 1981. All substances that are listed in the EINECS can be defined as phase-in substances. For a Non-EU manufacturer all substances that he is manufacturing and that are not listed in the EINECS are non-phase-in substances.

For a manufacturer within the EU in accordance with REACH Article 3 (20) [11] can be found some more substances that could have phase-in status as the legal text defines: "Phase-in substance: means a substance which meets at least one of the following criteria:

a) It is listed in the European Inventory of Existing Commercial Chemical Substances (EINECS);

b) It was manufactured in the Community, or in the countries acceding to the European Union on 1 January 1995 or on 1 May 2004, but not placed on the market by the manufacturer or importer, at least once in the 15 years before the entry into force of this Regulation, provided the manufacturer or importer has documentary evidence on this;

c) It was placed on the market in the Community, or in the countries acceding to the European Union on 1 January 1995 or on 1 May 2004, before entry into force of this Regulation by the manufacturer or importer and was considered as having been notified in accordance with the first indent of Article 8(1) of Directive 67/548/EEC but does not meet the definition of a polymer as set out in this Regulation, provided the manufacturer or importer has documentary evidence on this;" [11]

The substances mentioned in Article 3 (20) (b) [11] will be on-site isolated intermediates in case of manufacturers and in case of importers we can reckon on transported isolated intermediates.

When a company is located within the EU it was allowed for them in the time of pre-registration (until November 2008) to claim the phase-in status for such a substance. In any case, the phase-in status of such a substance is only valid for the claiming legal entity not for others that cannot benefit from Article 3 (20) (b) [11].

3.5
No-Longer Polymers

NLP is the abbreviation for No-Longer polymer. Under Directive 67/548/EEC polymers were subject to special rules [12]. This rules were completely different from the rules that have to be applied under REACH. No-longer polymers are substances that were considered to be polymers under the rules for EINECS [10] and therefore were not included in EINECS. As the term polymer was further defined in the 7th amendment of Directive 92/32/EEC some substances were no longer considered to be polymers [12]. Industry was then asked to submit their candidates [12]. These substances are now listed in the NLP-list [12] and have EC numbers with the format 5xx-xxx-x. The list is not an exhaustive list [12], but substances included in this list will be regarded as no-longer polymers by the authorities.

The NLP list consists mainly of the following groups: alkoxylated substances, oligomeric reaction products, oligomers from one monomer only, dimers and trimers, polymer-like substances containing 50% or more by weight of species with the same molecular weight.

Under REACH, substances with EC numbers 5xx-xxx-x can be treated as phase-in substances.

3.6
Non-Phase-In Substances

In general, all substances that do not fall under the definition phase-in substance are non-phase-in substances.

As there are different procedures for the registration process and different registration deadlines for phase-in substances and non-phase-in substances it is important to make clear what sort of substance you will have to register. It is highly recommended to check at first whether there is an entry for your substance in EINECS or not.

When you do not find an entry in EINECS nor the NLP list for a certain substance and you were not in the position to claim a phase-in status for your company in 2008 and the substance was not notified before REACH entered into force, you have to face the fact that it is a non-phase-in substance you have to deal with.

Consequences concerning registration deadlines are mentioned in Section 4.2 of this book.

3.7
Substances that Already Have Been Notified

Substances that were not listed in EINECS [10] and that were not regarded as polymers had to be notified under Directive 67/548/EEC [13].

REACH Article 3 (21) defines: "Notified substance: means a substance for which a notification has been submitted and which could be placed on the market in accordance with Directive 67/548/EEC;" [13]. These former notified new substances (NONS) are regarded as registered under REACH for the former notifier, but not for other companies. The former notifier has the right to claim a registration number for a notified substance. This process is described in detail in an Industry User Manual [14].

In any case, the claim of the registration number for NONS is a prerequisite for doing updates at a later stage. Dossiers of NONS should have been updated with the CLP classification without undue delay after 01 December 2010, as to do so the former notifier needed the corresponding IUCLID file from the national authorities. In some cases the national authorities had to be asked for the files and therefore there may have been some delay in doing the requested updates. ECHA therefore recommended industry to submit the updates as soon as practicable.

For the update of registrations of former notified substances there can be applied special rules as set out in the corresponding Data Submission Manual [15]. We will see some of the most important aspects in Section 11.4 of this book.

3.8
Overview on Official EC Numbers and not Official List Numbers

It is important to know what sort of substance you deal with before further obligations that your company may have in manufacturing or importing a certain substance can be defined and further actions can be planned accurately.

The list in Figure 3.1 may help you whenever you check the situation for a substance manufactured or used by your company and for that a procedure shall be defined on how to achieve REACH compliance. To find out whether you deal with a phase-in substance, a NLP or a non-phase-in substance you have to check

	Format of EC/List number	Source
Official (EC number)	2xx-xxx-x	EINECS List (European Inventory of Existing Commercial Chemical Substances)
	3xx-xxx-x	EINECS List (European Inventory of Existing Commercial Chemical Substances)
	4xx-xxx-x	ELINCS List (European List of Notified Chemical Substances)
	5xx-xxx-x	NLP List (No-Longer Polymers List)
Not official (List number)	6xx-xxx-x	Assigned to pre-registrations of substances with CAS number, not having an EC number yet
	7xx-xxx-x	Assigned to non-phase-in substances as a result of an Inquiry
	9xx-xxx-x	Assigned to pre-registrations for substances without a CAS number or any other numerical identifiers

Figure 3.1 Format of EC numbers and list numbers.

the format of the EC number of the corresponding substance. After having checked the list below [16] further To Dos can be derived by using the Check-List for Business Managers in Section 3.10 of this book.

Be aware of the fact that list numbers do not have any legal significance, but they are necessary as technical identifiers for processing a submission via REACH-IT. After registration of a substance that prior (in the pre-registration process or after an Inquiry) received a list number, your company will receive an official EC number. For future updates the official EC number will be required as the former list number in the EC field of the updated registration dossier will lead to Business rule failure in trying to submit the dossier to ECHA. In such a case the dossier cannot be further processed within ECHA and you will have to submit the dossier again after amending the list number into the official EC number.

3.9
Exemptions from REACH

There are some exemptions concerning REACH. Either a substance is exempted, because of special circumstances in production of the substance (the substance is not isolated or it is manufactured only in small amounts, e.g., lab scale) or there are some other regulations or directives under which a substance can fall because of a certain use.

An overview on exemptions is given in Article 2 of the REACH regulation [17] and also in Annex IV [18] and Annex V [19].

3.9.1
Non-isolated Intermediates

As already reported in Section 3.3.2.1 non-isolated intermediates do not have to be considered under REACH – they are exempted from REACH registration obligations. Please keep in mind that when you change a manufacturing process within your company and afterwards such a substance shall be isolated, there will occur registration obligations immediately at least for an on-site isolated intermediate.

3.9.2
Substances Manufactured or Imported in Amounts below 1 t/a

When a manufacturer located within the EU manufactures a substance in an amount of less than 1 t/a or an importer within the EU imports a certain substance in amounts of less than 1 t/a, there are no registration obligations to be fulfilled.

REACH is more generous than the rules that had to be applied in former days concerning the amount of new substances, that had to be notified in amounts from 10 kg/a and above.

Please be aware that an importer located within the EU has to sum up the amounts of a certain substance that he purchases from different Non-EU suppliers that do not take care of the registration obligations.

If an Importer located in Germany buys 200 kg of a substance A from a Chinese manufacturer in the end of 2013 and another 800 kg from an Indian supplier in the beginning of 2014, he does not have any obligation to register. If he buys from both suppliers within one calendar year, the amounts of both suppliers have to be added and in total the importer bought 1000 kg of substance A in the calendar year – then the importer suddenly will have registration obligations.

3.9.3
Substances Mentioned in Annex IV

In accordance with REACH Article 2 (7) (a) [20] in Annex IV of the REACH regulation [18] there are listed substances for which enough information was available to assume that their risks to human health and the environment are already well known. The list of exempted substances includes among others water and nitrogen. As air consists of nitrogen (ca. 78%) and oxygen (ca. 21%), the release of nitrogen into the environment will normally not lead to a high risk for human health, although it is known that already small deviations concerning the composition of air may have a great influence on human beings. Mountain climbers know the effects that a slightly reduced content of oxygen can have on them in great height compared to sea level. Liquefied nitrogen also may cause harmful effects if it is put on the skin. In medicine it is used to remove warts. However, it may be assumed that substances listed in Annex IV [18] cause minimum risk because of their intrinsic properties [18]. This can be applied also on the other substances that are listed in Annex IV [18].

3.9.4
Substances Listed in Annex V

REACH Article 2 (7) (b) [21] defines that certain substances are exempted from REACH registration obligations and they are "covered by Annex V, as registration is deemed inappropriate or unnecessary for these substances and their exemption . . . does not prejudice the objectives of this Regulation;" [21]

In Annex V [19] there are listed substances that result from chemical reactions that occur incidental during storage [22] of a certain substance for example, by environmental factors such as air, moisture, microbial organisms or sunlight [23]. Exempted are also substances that occur as a consequence of the end use of a certain substance [24]. Very important is the fact that by-products are exempted unless they are imported or placed on the market themselves [25]. When a company manufactures a certain substance and there is isolated sodium chloride as a by-product it is an important question what to do with this sodium chloride. If it is defined as waste and maybe is incinerated because national regulations allow to do so, no registration obligations will occur. If the company itself is located in the EU and will use the sodium chloride itself for example, in an electrolysis to manufacture chlorine, they have to do at least a registration for an on-site isolated intermediate. If the sodium chloride is sold to another company and there will be used as an intermediate at least a registration as a transported isolated intermediate will be required. If the sodium chloride shall be used in winter to put it on the streets to prevent the occurrence of glazed frost it is an end use and therefore a full registration will be required for a substance that in principle has the same composition as the material received from a salt-mine. However, the situation will be different in so far as the sodium chloride from the natural source in accordance with Annex V (8) [26] is exempted from the registration obligations, because it is not chemically modified and does not meet the criteria for classification as dangerous according to Directive 67/548/EEC [26], whereas the company manufacturing sodium chloride as a by-product has to register in case the by-product shall be used or marketed. Maybe because of the high costs and the workload that have to be considered for the preparation of a registration dossier and submission to ECHA, a company may decide to avoid using/marketing the sodium chloride that is received as a by-product. In accordance with Annex V (9) there are also exempted basic elemental substances for which hazards and risks are already well known: hydrogen, oxygen, noble gases (argon, helium, neon, xenon), nitrogen [27]. Nitrogen is exempted also in accordance with Annex IV [18], as we saw in Section 3.9.3.

3.9.5
Substances in the Interest of Defense

Member States may allow for exemptions from REACH in specific cases for certain substances, on their own, in a preparation or in an article, where necessary in the interests of defense [28]. Hopefully, no Member State will be interested in making use of this section in future.

3.9.6
Waste and Recovered Substances

As REACH Article 2 (2) [29] states: "Waste as defined in Directive 2006/12/EC of the European Parliament and of the Council is not a substance, preparation or article within the meaning of Article 3 of this Regulation" [29], there are no registration obligations for waste.

When a registered substance is recovered and afterwards used again or sold to another party it may be exempted from a further registration obligation under certain conditions. If for example, company A uses sulfuric acid (99%, 1% water) from a European manufacturer M for drying of a gas that is let through this sulfuric acid and then the resulting sulfuric acid (98%, containing 2% of water) is recovered and sold to another company B there will be no obligation to register the recovered sulfuric acid. Company A is not obliged to present a registration number in the SDS given to their customer B as the sulfuric acid is recovered material and will be identical to the already registered substance. If they intend to provide a registration number the last four digits of the registrant's registration number shall be omitted to be in compliance with the competition law.

3.9.7
Polymers

Owing to the potentially extensive number of different polymer substances on the market, and since polymer molecules are generally regarded as representing a low concern due to their high molecular weight, this group of substances is exempted from registration and evaluation under REACH [30].

"Polymer: means a substance consisting of molecules characterized by the sequence of one or more types of monomer units. Such molecules must be distributed over a range of molecular weights wherein differences in the molecular weight are primarily attributed to differences in the number of monomer units" [31]. "Monomer: means a substance which is capable of forming covalent bonds with a sequence of additional like or unlike molecules under the conditions of the relevant polymer-forming reaction used for the particular process" [32].

As a polymer in the meaning of REACH is exempted from registration obligations in accordance with Article 2 (9) [33], but instead of the polymer the corresponding monomer(s) are subject of registration obligation(s) it is important to define the term polymer more in depth. A polymer comprises the following [31]:

a) A single weight majority of molecules containing at least three monomer units that are covalently bound to at least one other monomer unit or other reactant [34]. In other words: Over 50% of the weight of that substance consists of polymer molecules [30].

b) Less than a simple weight majority of molecules of the same molecular weight [35]. In other words: The amount of polymer molecules representing the same molecular weight must be less than 50 weight percent of the substance [30].

A "monomer unit" means the reacted form of a monomer substance in a polymer [31]. This means a monomer is in the polymerization reaction converted into a repeating unit of the polymer sequence [30]. Any monomer is therefore an intermediate, but the specific provisions for the registration of intermediates under REACH do not apply to monomers [30, 36]. For applications outside the scope of polymerization such a substance may be regarded as a transported isolated intermediate or an on-site isolated intermediate, if it fulfills the conditions as set out in Article 18 (4) [9] or Article 17 (3) [7] of the REACH regulation.

Substances exclusively involved in the catalysis, initiation or termination of the polymer reaction are not monomers [30].

As stated in Article 6 (3) [37] "Any manufacturer or importer of a polymer shall submit a registration to the Agency for the monomer substance(s) or any other substance(s), that have not already been registered by an actor up the supply chain, if both the following conditions are met [37]:

a) The polymer consists of 2% weight by weight (w/w) or more of such monomer substance(s) or other substance(s) in the form of monomeric units and chemically bound substance(s) [38].

b) The total quantity of such monomer substance(s) or other substance(s) makes up 1 ton or more per year [39]."

As it may be a great challenge to make a decision in every single case whether a substance is a polymer or not and further steps in defining registration obligations and tasks are based on this fundamental decision there are two examples given in Figure 3.2 [40] that may assist you to gain some experience in this field.

How to find out whether a substance can be considered as a polymer under REACH or not

Example 1		Example 2	
R-M	24%	R-M	9%
R-M-M	34%	R-M-M	17%
R-M-M-M	19%	R-M-M-M	30%
R-M-M-M-M	11%	R-M-M-M-M	23%
R-M-M-M-M-M	7%	R-M-M-M-M-M	15%
R-M-M-M-M-M-M	3%	R-M-M-M-M-M-M	5%
R-M-M-M-M-M-M-M	2%	R-M-M-M-M-M-M-M	3%
Rule 1: No chain $\geq$ 50% → YES ✓		Rule 1: No chain $\geq$ 50% → YES ✓	
Rule 2: (3M + 1)-rule (More than 50% of the molecules comprise of at least 3 monomers bound covalently to another reactant). 19% + 11% + 7% + 3% + 2% = 42% < 50% → NO Polymer		Rule 2: (3M + 1)-rule: 30% + 23% + 15% + 5% + 3% = 72% > 50% → YES ✓	
Substance in Example 1 is not a polymer in the meaning of REACH		Substance in Example 2 is a polymer in the meaning of REACH	

R = Other reactant
M = Monomer

Figure 3.2 Polymer in the meaning of REACH.

3.9.8
Re-imported Substances

If an already registered substance is exported from the Community by an actor in the supply chain and at a later stage is re-imported into the Community by the same or another actor in the supply chain, this substance may be exempted from further registration obligations. The re-importer has to show that

1) the substance being re-imported is the same as the exported substance [41];
2) he has been provided with the information in accordance with Article 31 or 32 relating to the exported substance [41].

At first it may sound strange to export a substance from the EU to a Non-EU country and afterwards re-import the same substance again. In daily life there may be good reasons for acting in that way. If for example, a Swiss company purchases a solvent from a German manufacturer/importer this solvent will be registered by the German manufacturer/importer. The Swiss company may use the mentioned solvent as such in a manufacturing process for a certain substance. If the manufactured substance as such will be sold to customers, nobody will be interested in whether the Swiss company used a solvent that had been registered. If the manufactured substance shall be sold as a solution in the mentioned solvent to a EU customer, it will be a requirement to use a registered solvent as normally the customer is not willing to register a solvent in the role of the importer when such a solvent is easily available. The Swiss company also may make the decision that it is more comfortable to buy the solvent from a European source than to take care of the registration obligations via an Only Representative when the price for the solvent from the European source is competitive.

However, when the Swiss company also sells the solution to Non-EU customers it may be considered to use for these customers a solvent that was not registered under REACH as it may be cheaper.

3.9.9
Further Exemptions: Use in Medicinal Products or for Food and Feedingstuffs

There are no registration obligations under REACH when a substance is manufactured exclusively for the use in medicinal products within the scope of Regulation (EC) No726/2004, Directive 2001/82/EC of the European Parliament and of the Council of 6 November 2001 on the Community code relating to veterinary medicinal products and Directive 2001/83/EC of the European Parliament and of the Council of 6 November 2001 on the Community code relating to medicinal products for human use [42].

In accordance with the exemption from the registration obligations for the mentioned substances, preparations in the finished state, intended for the final use within the scope of the above-mentioned laws are exempted from the obligations concerning "Information in the supply chain" described in Title IV [43, 44].

The exemption from registration is applicable to substances manufactured within the EU and afterwards used either in the EU or exported to Non-EU countries. Imports of substances for that use are also covered [45].

The exemption does not distinguish between active or nonactive ingredients as it applies to any substance "used in medicinal products" [45]. Excipients used in medicinal products are therefore also exempted from registration [45].

In accordance with Article 2 (5) (b) [46] the use of a substance in food or feedingstuffs in accordance with Regulation (EC) No 178/2002 including use: as food additive in foodstuffs [47], flavoring in foodstuffs [48], additive in feedingstuffs [49] or in animal nutrition [50] is also exempted from registration obligations [46]. A reason for this exemption is that "the Food Safety Regulation already requires that food for humans cannot be placed on the market unless it is, that is, not injurious to human health and fit for human consumption [45]."

3.9.10
Product and Process Orientated Research and Development

Since one of the main objectives of REACH is to enhance innovation [45], there are special rules for substances manufactured and used in the field of research and development. Scientific research and development as defined in Article 3 (23) [51] means any scientific experimentation, analysis or chemical research carried out under controlled conditions in a volume less than 1 ton per year [51]; a substance being used solely for scientific research and development does not need to be registered since the registration obligation applies to volumes of one tone or more per year [45].

Product and process orientated research and development (PPORD) is defined by Article 3 (22) [52]. This means any scientific development related to product development or the further development of a substance, on its own, in preparations or in articles in the course of which pilot-plant or production trials are used to develop the production process and/or to test the fields of application of the substance [52]; in general there will be manufactured amounts above 1 t/a, but as demanded by Article 9 (1) [53] quantity is limited to the purpose of product and process orientated research. However, substances used for PPORD in quantities of one ton or more per year will receive an exemption from registration for five years if they are notified to ECHA [45]. ECHA may extend the exemption period for up to a further five years (or ten years in the case of medicinal products or substances not placed on the market) upon request, as long as this can be justified by the program of research and development presented by the notifier [45].

Details concerning conditions and data requirements for PPORD notifications will be discussed in Section 6.1 of this book.

3.9.11
Substances Regarded as Being Registered

In REACH Article 2 [54] a description of the field of application of the regulation can be found. In addition, there are listed cases in which a substance does not have to be registered under REACH, because they are exempted in general or another law has to be applied.

In the previous sections situations were described when substances could be exempted from REACH registration obligations because of special properties or certain circumstances. Now there are left few cases in which substances are exempted from REACH registration obligations, because they are regarded as registered under REACH.

These substances are defined in Article 15 [55] as "Substances in plant protection and biocidal products" and in Article 24 (1) [56] as "Notified Substances". Duties of the Commission, the Agency and registrants of substances regarded as being registered are laid down in Article 16 [57].

Active substances and co-formulants manufactured or imported for use in plant-protection products only shall be regarded as being registered and the registration as completed for manufacture or import for the use as a plant protection product when it is included in at least one of the following [58]:

- Annex I to Directive 91/414/EEC;
- Regulation (EEC) No 3600/92;
- Regulation (EC) No 703/2001;
- Regulation (EC) No 1490/2002;
- Decision 2003/565/EC;
- Commission decision on the completeness of the dossier has been taken pursuant to Article 6 of Directive 91/414/EEC.

Please note that quantities of the same active substance used for uses other than in plant protection products are not regarded as being registered even if they are included in one of the above-mentioned documents [45]. This means only active substances can be regarded as registered, whereas other substances for example, co-formulants used for producing the plant protection product need to be registered [45].

Active substances manufactured or imported for use in biocidal products only shall be regarded as being registered and the registration as completed for manufacture or import for the use in a biocidal product when it is included in at least one of the following [59]:

- Annexes I, IA or IB to Directive 98/8/EC of the European Parliament and of the Council of 16 February 1998 concerning the placing of biocidal products on the market [59];

- Commission Regulation (EC) No 2032/2003 of 4 November 2003 on the second phase of the 10-year work program referred to in Article 16 (2) of Directive 98/8/EC, until the date of the decision referred to in the second subparagraph of Article 16 (2) of Directive 98/8/EC [59].

Please note that only active substances can be regarded as registered, whereas other substances used for producing the biocidal product are subject to registration under REACH [45].

The third category of substances that can be regarded as registered are notified substances, as a notification in accordance with Directive 67/548/EEC shall be regarded as a registration [56]. As the former NONS were already considered in Section 3.7 here they are only mentioned for the sake of completeness.

3.9.12
How to Cope with Situations in Which Parts of the Manufactured Amount are Falling under REACH and Another Part is Exempted

Situations in which a part of a manufactured substance has to be registered under REACH and another part of the total amount that is manufactured by a company is exempted from REACH are difficult. In any case, it will be a great challenge to document the situation accurately within the company.

There is no obligation to use different product numbers and especially when the quality of the product is the same, although the use will be different at a later stage it may seem unnecessary to differentiate. However, it is a great challenge to solve all problems that can occur within the supply chain without doing so. In many cases a company will have to provide different SDS in accordance with the special situation, for example, for a substance that is used in medicinal products and therefore the corresponding amount is exempted from REACH the customer should not receive a SDS where the registration number is included, whereas a customer that uses the mentioned substance as an intermediate in the synthesis of another substance should receive a SDS containing a registration number. When part of the substance is registered as a transported isolated intermediate and another part is registered with end uses (full registration) it has to be ensured that customers using the product with end use will receive an eSDS, whereas customers that use the product as transported isolated intermediate (TII) and therefore confirmed the use under strictly controlled conditions in accordance with Article 18 (4) [9] shall not receive an eSDS.

Within the supply chain of the company that manufactures a substance for different purposes and has therefore to differentiate it is highly recommended to implement the use of several product identifiers within the company for example, product number 000001-00 for a product 1 as it is manufactured, product number 000001-01 for product 1 that is later used as an on-site isolated intermediate, product number 000001-02 for the amount that is registered as a transported isolated intermediate, product number 000001-03 for the amount that is registered with end use and may be product number 000001-04 for the amount of the product 1 that is exempted from REACH registration obligations because of a certain reason. Only by differentiating in this or a similar way can it be ensured that it is possible to trace back to the type of use that a certain part of the total amount manufactured within a calendar year in your company was used by the customer(s) and to have proof in case of on-site inspections of national authorities at your site.

When your company works with different product numbers it will also be possible to ensure that the right version of the SDS is distributed to each of your customers by the support of an IT-tool (e.g., the widely used SAP).

3.10
Check-List for Business Managers

This list can be useful for Business Mangers especially in the evaluation of a new project in the field of Custom Manufacturing.

Substance name (Abbreviation)	CAS	EC
1) Decision – step1 EC 2xx-xxx-x → phase-in substance, go on with 2a) EC 3xx-xxx-x → phase-in substance, go on with 2a) EC 5xx-xxx-x → NLP (quasi phase-in), go on with 2a) EC 4xx-xxx-x → non-phase-in substance, go on with 2b) No EC available → non-phase-in substance, go on with 2b) Polymer in the meaning of REACH → check situation for monomer		Structural formula
2) Decision – step 2		
a) Phase-in substances (including NLP)	b) Non-phase-in substances	
• CMR Cat. 1 or 2? ☐ yes ☐ no • R50/53? ☐ yes ☐ no → ask HSE/ESH department		
• Tonnage band ☐ below 1 t/a → no REACH relevance ☐ 1 to <10 t/a ☐ 10 to <100 t/a ☐ 100 to <1000 t/a ☐ >1000 t/a		

→Registration deadline
☐ 30.11.2010 for >1000 t/a
    R50/53 and >100 t/a
    CMR Cat. 1or 2 and >1 t/a
☐ 31.05.2013 for 100 to <1000 t/a
☐ 31.05.2018 for 10 to <100 t/a
    1 to 10 t/a
	→Registration deadline
	Before manufacturing or import of 1 t/a and above (Inquiry Dossier, Registration Dossier)	

Is (late) pre-registration still possible? ☐ yes ☐ no → Procedure as for non-phase-in substances (Inquiry Dossier, Registration Dossier)		
3) Decision – step 3 Type of registration ☐ full registration (use of substance as solvent, catalyst, monomer for production of polymer, any other end use, intermediates – if not manufactured and used under strictly controlled conditions) ☐ transported isolated intermediate (TII) ☐ on-site isolated intermediate (OII) ☐ non-isolated intermediate → no registration required under REACH		

(Continued)

3 What Sort of Substances have to be Considered under REACH

Substance name (Abbreviation)	CAS	EC
4) Decision – step 4 Data requirements concerning studies/tests ☐ OII and TII<1000 t/a → no special data requirements ☐ TII>1000 t/a → REACH Annex VII ☐ full 1 to 10 t/a → REACH Annex VII ☐ full 10 to 100 t/a → REACH Annex VII, Annex VIII and additional CSR ☐ full 100 to 1000 t/a → REACH Annex VII, VIII and IX and additional CSR ☐ full >1000 t/a → REACH Annex VII, VIII, IX and X and additional CSR		
5) Next steps to be done by the REACH Department ☐ (late) pre-registration ☐ prepare Inquiry Dossier ☐ do cost estimation	Remarks	

3.11
Examples and Exercises

1. In the following list you will find substances, preparations and articles in the meaning of REACH. Please decide in each case whether the given product is a substance, a preparation or an article and give reasons for your decision.

Product	Substance, preparation or article	Reasons
Benzene		
Bottle		
Chair		
Chlorine		
Dodecanediocic acid		
Hydrogen chloride		
Hydrochloric acid		
Iodine		
Methanol		
Mixture of methanol and ethanol		
Oleum		
Plastic knife		
Sodium chloride		
Solution of iodine in benzene		
Sulfuric acid (100%)		
Sulfur trioxide		
Tetrahydrofurane		
Vacuum cleaner		
Xanthydrol (10% in Methanol)		

2. Below is given the composition of several substances as they are manufactured and afterwards brought to market. Decide for each substance whether a mono-constituent substance, a multi-constituent substance or an UVCB has to be mentioned in the registration dossier.

   a) Sulfuric acid (98%)

Composition	Proportion (w/w)
Sulfuric acid	98%
Water	2%

   b) Sulfuric acid (70%)

Composition	Proportion (w/w)
Sulfuric acid	70%
Water	30%

   c) A Fragrance consists of Substance S and its isomers. It is manufactured in two different qualities (Quality Excellent and Quality Normal).

Fragrance	Quality excellent	Quality normal
Composition	Proportion (w/w)	Proportion (w/w)
Substance S	80–85%	65–75%
Isomer 1	6–9%	3–8%
Isomer 2	3–10%	10–21%
Isomer 3	0.5–1.5%	2–5%
Isomer 4	0.5–1.5%	4–6%
Isomer 5	0.5–1.5%	1–4%

   d) Hydrochloric acid (37%)

Composition	Proportion (w/w)
Hydrogen chloride	37%
Water	63%

3. In the list below you will find substances and their CAS number. Please find out for each substance whether it is a phase-in substance, a NLP or a non-phase-in substance and give a reason for your decision.

Substance	CAS	Phase-in substance	NLP	Non-phase-in substance
2-methylisocrotonic acid (Angelic acid)	565-63-9			
Angelic acid methyl ester	5953-76-4			
Angelic Anhydride	94487-74-8			
2-Bromopropionyl bromide	563-76-8			
Benzyl alcohol	100-51-6			
D-(+)-Turanose	547-25-1			
Biotin	22879-79-4			
n-Valeryl bromide	1889-26-5			
n-Valeryl chloride	638-29-9			
Acetic acid	64-19-7			
Sulfuric acid	7664-93-9			
Butene, dimers	9021-92-5			
Hexamethylene diisocyanate, oligomers	28182-81-2			
4-amino-6-[(1,2-dihydro-2-imino-1,6-dimethyl-4-pyrimidinyl)amino]-1,2-dimethylquinolinium hydroxide	93919-18-7			
Phenol, propoxylated	28212-40-0			
2,2'-[(4-[2-hydroxyethyl)amino]-2-nitrophenyl]imino] bisethanol	93919-22-3			
Aluminum isopropoxide	555-31-7			
Oleic acid, ethoxylated	9004-96-0			
Vanillin	121-33-5			
Xanthydrol	90-46-0			

4. Under which circumstances was your company allowed to pre-register the substance Angelic Anhydride (mentioned in Question number 3)?

5. Nitrogen has CAS (7727-37-9) and EC (231-783-9) – Was it possible to pre-register this substance in 2008? Why?

6. Hydrogen is produced in your company as a by-product in a certain process.

   a) Are you allowed to market this substance? Please give a reason for your statement.

   b) Is the situation different for other by-products that are marketed? Please give a reason for your statement.

c) Are there any registration obligations in case your company makes the decision not to bring to market such a by-product and instead either uses the substance or it is considered to be waste and is burned.

References

1 Regulation (EC) No 1907/2006 of the European Parliament and of the Council of 18 December 2006, concerning the Registration, Evaluation, Authorisation and Restriction of Chemicals (REACH), establishing a European Chemicals Agency, amending Directive 1999/45/EC and repealing Council Regulation (EEC) No 793/93 and Commission Regulation (EC) No 1488/94 as well as Council Directive 76/769/EEC and Commission Directives 91/155/EEC, 93/67/EEC, 93/105/EC and 2000/21/EC.
2 See Article 3 (1) of the REACH regulation [1].
3 See Article 3 (2) of the REACH regulation [1].
4 See Article 3 (3) of the REACH regulation [1].
5 Guidance for identification and naming of substances under REACH and CLP, Version: 1.1, November 2011, published by the European Chemicals Agency.
6 See Article 3 (15) of the REACH regulation [1].
7 See Article 17 (3) of the REACH regulation [1].
8 Guidance on intermediates, Version: 2, December 2010, published by the European Chemicals Agency; Guidance documents can be obtained via the website of the European Chemicals Agency (http://echa.europa.eu/reach_en.asp).
9 See Article 18 (4) of the REACH regulation [1].
10 The online EINECS Information system enables you to find, through the European INventory of Existing Commercial chemical Substances general information concerning a substance like CAS number, EINECS number, substance name and Chemical Formula; EINECS published in O.J. C 146 A, 15.6.1990; EINECS corrections published in O.J. C 54/13 01.03.2002, 2002/C54/08. http://esis.jrc.ec.europa.eu/index.php?PGM=ein
11 See Article 3 (20) of the REACH regulation [1].
12 Notification of new chemical substances in accordance with Directive 67/548/EEC on the classification, packaging and labeling of dangerous substances, NO-LONGER POLYMER LIST, Version 3, published by the European Chemicals Bureau.
13 Council Directive of 27 June 1967 on the approximation of laws, regulations and administrative provisions relating to the classification, packaging and labelling of dangerous substances (67/548/EEC).
14 REACH-IT Industry User Manual, Part 10 – Claim of a registration number for a notified substance, Version 2.0, July 2012, published by the European Chemicals Agency, http://echa.europa.eu/
15 See Annex 4 in the Data Submission Manual Part 05 – How to complete a technical dossier for registrations and PPORD notifications, Version 3.0, 07/2012, published by the European Chemicals Agency, http://echa.europa.eu/
16 Guidance on data sharing, Version 2.0, April 2012, published by the European Chemicals Agency, p. 30. Guidance documents can be obtained via the website of the European Chemicals Agency (http://echa.europa.eu/reach_en.asp).
17 See Article 2 of the REACH regulation [1].
18 See Annex IV of the REACH regulation [1].
19 See Annex V of the REACH regulation [1].
20 See Article 2 (7) (a) of the REACH regulation [1].
21 See Article 2 (7) (b) of the REACH regulation [1].

22 See Annex V (2) of the REACH regulation [1].
23 See Annex V (1) of the REACH regulation [1].
24 See Annex V (3) of the REACH regulation [1].
25 See Annex V (5) of the REACH regulation [1].
26 See Annex V (8) of the REACH regulation [1].
27 See Annex V (9) of the REACH regulation [1].
28 See Article 2 (3) of the REACH regulation [1].
29 See Article 2 (2) of the REACH regulation [1].
30 Guidance for monomers and polymers, Version 1.1, May 2008, published by the European Chemicals Agency.
31 See Article 3 (5) of the REACH regulation [1].
32 See Article 3 (6) of the REACH regulation [1].
33 See Article 2 (9) of the REACH regulation [1].
34 See Article 3 (5) (a) of the REACH regulation [1].
35 See Article 3 (5) (b) of the REACH regulation [1].
36 See Article 6 (2) of the REACH regulation [1].
37 See Article 6 (3) of the REACH regulation [1].
38 See Article 6 (3) (a) of the REACH regulation [1].
39 See Article 6 (3) (b) of the REACH regulation [1].
40 For further assistance see also the document "Besonderheiten bei Polymeren und Monomeren", 3. überarbeitete Auflage, June 2010, published by the Bundesanstalt für Arbeitsschutz und Arbeitsmedizin (BAuA), Dortmund (Germany), ISBN 978-3-88261-674-3.
41 See Article 2 (7) (c) of the REACH regulation [1].
42 See Article 2 (5) (a) of the REACH regulation [1].
43 See Titele IV Information in the supply chain, comprising Article 31 – Article 36 of the REACH regulation [1].
44 See Article 2 (6) (a) of the REACH regulation [1].
45 Guidance on registration, Version 2.0, May 2012, published by the European Chemicals Agency.
46 See Article 2 (5) (b) of the REACH regulation [1].
47 Within the scope of Council Directive 89/107/EEC of 21 December 1988 on the approximation of the laws of the Member States concerning food additives authorized for use in foodstuffs intended for human consumption.
48 Within the scope of Council Directive 88/388/EEC of 22 June 1988 on the approximation of the laws of the Member States relating to flavourings for use in foodstuffs and to source materials for their production and Commission decision 1999/217/EC of 23 February 1999 adopting a register of flavouring substances used in or on foodstuffs drawn up in application of Regulation (EC) No 2232/96 of the European Parliament and of the Council.
49 Within the scope of Regulation (EC) No 1831/2003 of the European Parliament and of the Council of 22 September 2003 on additives for use in animal nutrition.
50 Within the scope of Council Directive 82/471/EEC of 30 June 1982 concerning certain products used in animal nutrition.
51 See Article 3 (23) of the REACH regulation [1].
52 See Article 3 (22) of the REACH regulation [1].
53 See Article 9 (1) of the REACH regulation [1].
54 See Article 2 of the REACH regulation [1].
55 See Article 15 of the REACH regulation [1].
56 See Article 24 (1) of the REACH regulation [1].
57 See Article 16 of the REACH regulation [1].
58 See Article 15 (1) of the REACH regulation [1].
59 See Article 15 (2) of the REACH regulation [1].

4
Obligation to Submit a Registration Dossier

4.1
Who has to Register? Who may Register?

Every manufacturer within the EU who produces a chemical substance in amounts of 1 t/a or more has the obligation to register said substance under REACH.

When a trader or a manufacturer imports a substance in amounts of 1 t/a or more from a Non-EU country, in general the importer has the obligation to register the mentioned substance under REACH. This obligation can only be negated if the Non-European manufacturer of the mentioned substance takes care of the registration obligations via an Only Representative who is located in the European Union and submits the registration dossier to ECHA.

Within globally acting enterprises very often the situation will occur that there are several separate legal entities located in different countries and may even be on different continents of the world. When one or more legal entities are located in the European Union it is possible that such a legal entity can be used as an Only Representative for the Non-European legal entities. If there is no other legal entity available that is located within the EU within a company group there can also be a consultant instructed to take over the role of the Only Representative. Asian companies that do not have associated companies within the EU but are interested in having a large number of customers within the EU very often are willing to pay for the service of the consultant and also the fees with ECHA, because European customers very often demand that their suppliers take over the registration obligations.

When a European manufacturer produces a substance that sometimes is also purchased from a Non-European manufacturer, it is possible to cover the substance manufactured from both manufacturers within one registration dossier. The European manufacturer registers this substance in the role of the Manufacturer and at the same time as an Importer. A different situation is given for Only Representatives. An Only Representative can represent a Non-EU manufacturer, but cannot have other rules within the same registration dossier. When an Only Representative represents more than one company concerning a certain substance, then for every company that he represents he must submit a registration

REACH Compliance – The Great Challenge for Globally Acting Enterprises, First Edition.
Susanne Kamptmann.
© 2014 Wiley-VCH Verlag GmbH & Co. KGaA. Published 2014 by Wiley-VCH Verlag GmbH & Co. KGaA.

	EU		**Non-EU**
Manufacturer (M)	Obligation to register substances manufactured in amounts of 1 t/a and above	Manufacturer (M)	No registration obligation, but may appoint an OR for the amounts of a substance that are delivered into the EU
Importer (I)	Obligation to register substances imported in amounts of 1 t/a and above		
Downstream User (DU)	Buys and uses substances that are already registered. In general no registration obligations		
Trader	May act either as an Importer or a "Downstream User"	Trader	Is not allowed to take care of registration obligations, is not allowed to appoint an OR. His EU-customers will be importers under REACH
Only Representative (OR)	May act as an "Importer" and take care of the registration obligations on behalf of a Non-EU Manufacturer		

Figure 4.1 Roles under REACH and resulting registration obligations.

dossier to ECHA. Therefore, the Only Representative may need access to more than only one account in REACH-IT.

For a short overview on different roles under REACH and the resulting registration obligations see Figure 4.1.

Within globally acting enterprises maybe it is intended to manufacture a certain substance at a Non-EU site and also to import this substance from another Non-European country. In this case a decision maker within the globally acting enterprise is recommended to think this over carefully. There have to be considered more than only REACH aspects and maybe in the end REACH is not decisive, but it is clear that it is advantageous to keep in mind that there are several options to achieve legal compliance in matters of REACH for the company.

Supposing that the company group EXTRACLEVER manufactures substance A at a site located in Switzerland and also has sites located within Europe and the substance A could also be purchased from a Chinese manufacturer. First, we have

4.1 Who has to Register? Who may Register? | 51

EXTRACLEVER (Switzerland)	EXTRACLEVER (EU)	Manufacturer of Substance A outside of the EU (China)
• Manufacturer of Substance A	• is not involved at all	• Manufacturer of Substance A
• Imports Substance A from China		
• sells Substance A (either manufactured at Extraclever (Switzerland) or purchased from Chinese Manufacturer) to Non-EU customers		
→ No registration obligations under REACH		

Figure 4.2 All customers of EXTRACLEVER are located outside of the EU.

to check where the customers of EXTRACLEVER are located to then develop a strategy of how to cope best with the REACH registration obligations.

If all customers of the company EXTRACLEVER are located outside of the EU it could be useful to import the amounts of A purchased from the Chinese manufacturer to the Swiss legal entity of EXTRACLEVER and sell it afterwards to the customers outside of the EU (see also Figure 4.2). In this case no registration obligations will apply.

If all customers are located within the EU the amounts of A produced at the Swiss site of EXTRACLEVER must be registered and also the amounts of A purchased from the Chinese manufacturer (see also Figure 4.3). When the Chinese manufacturer takes care of the registration obligations via an Only Representative in the EU, the EXTRACLEVER group has only to think about the import of the amounts manufactured at the Swiss site. In principle, the customers of EXTRACLEVER (Switzerland) are importers in the sense of REACH and would have the registration obligations. If every customer in the role of the importer were to import less than 1 t/a then nobody would have registration obligations. If the customers of EXTRACLEVER (Switzerland) purchase more than 1 t/a of substance A and they demand that EXTRACLEVER takes care of the REACH registration, there are still two options to cover all the needs. Either EXTRACLEVER (Switzerland) does the registration via Only Representative EXTRACLEVER (located in the EU) or EXTRACLEVER (located in the EU) acts as an Importer and therefore does the registration in the role of the importer.

The advantage when the registration is done via OR will be that without any doubt EXTRACLEVER (Switzerland) is allowed to deliver the substance A directly to their customers. When EXTRACLEVER (located in the EU) acts as an importer it may be that there are several other consequences concerning clearance and taxes. On the other hand, EXTRACLEVER (located in the EU) on doing a registration as an importer will be allowed not only to purchase from EXTRACLEVER (Switzerland), but also from any other Non-EU manufacturer or even from Non-EU

EXTRACLEVER (Switzerland)	EXTRACLEVER (EU)	Manufacturer of Substance A outside of the EU (China)
• Manufacturer of Substance A (can be registered via OR of EXTRACLEVER (Switzerland))	• May act as an OR for EXTRACLEVER (Switzerland)	• Manufacturer of Substance A (has no registration obligations, but may appoint an OR within the EU)
• Imports Substance A from China (cannot be registered via OR of EXTRACLEVER (Switzerland))	• May (also) act as an OR for the Chinese Manufacturer	
• sells Substance A to European customers (situation needs to be analyzed in depth to find an appropriate solution concerning REACH issues)	• May act as an importer for substance A (registration dossier could cover imports from EXTRACLEVER (Switzerland) and also from the Chinese Manufacturer and any other Non-EU manufacturer)	
→ Registration obligations will occur – one or several registration dossiers may be necessary depending on who will register and which roles under REACH can be covered in the corresponding dossier		

Figure 4.3 At least some customers of EXTRACLEVER are located within the EU.

traders. (Remark: Non-EU traders are not allowed to register under REACH – no registration via OR is possible.)

If EXTRACLEVER (Switzerland) already did a registration via OR EXTRACLEVER (located in the EU) and there occurs an additional demand for the Substance A from further customers that shall be covered via imports from the Chinese Manufacturer, EXTRACLEVER (Switzerland) may sell the amounts that they manufactured at their own site to European customers, as it will be covered by the already existing registration. The amounts purchased from the Chinese Manufacturer are not covered by this registration, because EXTRACLEVER (Switzerland) acts in this case as a Non-EU Trader. If possible, EXTRACLEVER (Switzerland) may sell the amounts purchased from the Chinese Manufacturer only to customers outside of the EU. Then, there will not be any further registration obligations. When it is necessary to deliver also at least parts of the amount purchased from the Chinese Manufacturer to EU customers the EXTRACLEVER group may use a European site acting as an importer under REACH, but in any case there will be a further registration required.

We have to learn from this more or less simple example that it very often will not be possible to find the best solution to achieve REACH compliance within five minutes, but we can also see that when we consider a set of information maybe we will find a good solution that already considers needs that could arise in the future.

4.2
Pre-registration and Late Pre-registration

Before REACh entered into force only non-phase-in substances were subject to notification in accordance with Directive 67/548/EEC. As REACH for the first time brought registration obligations concerning all formerly known substances, in particular the substances listed in EINECS it was clear that industry would need some time to prepare all the registration dossiers. It was an aim of the REACH regulation to bring together all future registrants that would have to register the same substance in order to avoid unnecessary animal testing. The motto of ECHA is "one substance one registration". In consideration of the time that would be necessary to bring together companies that are on the way to registering the same substance, the REACH regulation offered the possibility to pre-register all substances that a manufacturer already produced or an importer already had imported when these substances were phase-in substances or had at least phase-in status.

Therefore, European manufacturers and importers and also the Only Representatives of Non-EU manufacturers willing to register had the possibility to pre-register the mentioned phase-in substances (including substances that had phase-in status for a certain company) until 30 November 2008. Afterwards, companies that had pre-registered a substance could benefit from an extension of time until the registration dossier had/has to be submitted to ECHA.

Perhaps a company made the decision to manufacture a phase-in substance after this deadline for pre-registrations for the first time. In such a case it is allowed to make a late pre-registration, if the manufacturer/importer of the mentioned substance does the late pre-registration within the first six months after first manufacturing or first import and there are at least twelve months until the registration deadline. In all other cases the registration dossier has to be submitted immediately.

For a short overview on pre-registration and late pre-registration see Figure 4.4, and in case you would like to check whether your company can still do a late pre-registration for a certain substance Figure 4.5 may be helpful.

Very often, importers of raw materials pre-registered these substances, because it was not clear whether their non-European suppliers were in the position and also willing to pre-register via an Only Representative in the EU. Therefore, ECHA received many more pre-registrations than they had expected before the deadline for pre-registration in November 2008.

As a pre-registration provides a view on all other pre-registrants for the same substance it cannot be out of the question that some black sheep among the consultants used this option pretending that they would pre-register in the role of the Only Representative and in reality tried to do market research or offered later on IT-services for SIEF communication.

For a company that pre-registered a substance, but ceases manufacturing or does not have registration obligations concerning imported raw materials, it is possible to "inactivate" the corresponding pre-registration at any time in REACH-IT. If the company restarts manufacturing of this substance or has the obligation to register

	Substances that can be considered	Deadline for doing pre-registration or late pre-registration
Pre-registration	Phase-in substances as listed in REACH Article 3 (20) *)	30. November 2008
Late pre-registration	Phase-in substances in accordance with REACH Article 3 (20) (a) *) when a company manufactures/imports such a substance for the first time after the deadline for pre-registration in 2008	within six months after first manufacture/import when there are at least twelve months until the registration deadline

Figure 4.4 Pre-registration and late pre-registration.
*) Can be applied also to No-longer polymers that are listed in the NLP-list.

Substance and relevant tonnage band	Registration deadline for (late) pre-registered substances	Is late pre-registration still possible?
• CMR from 1 t/a • R50/53 from 100 t/a • >1000 t/a	30.11.2010	Not possible any longer → Inquiry and immediate registration as for non-phase-in substances
• 100 to 1000 t/a	31.05.2013	Was possible until 31.05.2012. Not possible any longer → Inquiry and immediate registration as for non-phase-in substances
• 10 to 100 t/a • 1 to 10 t/a	31.05.2018	Possible until 31.05.2017

Figure 4.5 Check whether late pre-registration is possible for a phase-in substance that your company manufactures/imports for the first time.

the raw material because the Non-European supplier does not register the material, it is also possible to "reactivate" the pre-registration.

Amendments of tonnage band and the corresponding registration deadline can also be done at any time.

Actually doing a pre-registration or an amendment of the pre-registration will last only a few minutes if your company already has a REACH-IT account.

In this REACH-IT account you can check all the pre-registrations that so far have been done for your company.

Registration deadline for a non-phase-in substance	
Tonnage band	Registration deadline
Less than 1 t/a	No registration required
1 t/a and above	Before first manufacture or import of 1 t/a and above

Figure 4.6 Registration deadline for non-phase-in substances.

Registration deadline for a (late) pre-registered phase-in substance	
Substance and relevant tonnage band	Registration deadline for (late) pre-registered substances
• CMR from 1 t/a • R50/53 from 100 t/a • >1000 t/a	30.11.2010
• 100 to 1000 t/a	31.05.2013
• 10 to 100 t/a • 1 to 10 t/a	31.05.2018
• Less than 1 t/a	No registration required

Figure 4.7 Registration deadline for a (late) pre-registered substance.

4.3
When Does a Substance have to be Registered?

Non-phase-in substances have to be registered before a manufacturer within the EU produces or an importer imports the mentioned substance in amounts of 1 t/a or more. In Figure 4.6 you will find an overview concerning the registration deadline for non-phase-in substances depending on the tonnage band. The same can be applied for phase-in substances when a company has forgotten to pre-register and/or is not allowed to do a late pre-registration any longer.

For phase-in substances that have been pre-registered before the deadline in 2008 or for those that have made a late pre-registration in time, the registrant to be can benefit from extended registration deadlines as follows:

CMR substances cat. 1 or 2 (in accordance with Directive 67/548/EEC) in amounts of 1 t/a or more had to be registered until 30. November 2010 [1, 2].

Substances that are classified as R50/53 (in accordance with Directive 67/548/EEC) manufactured or imported in amounts of 100 t/a and above also had to be registered until 30 November 2010 [3].

All other phase-in substances that had been pre-registered or late pre-registered have registration deadlines according to the tonnage band.

For the tonnage band >1000 t/a also 30 November 2010 had been the relevant registration deadline [4]. For the tonnage band 100 to 1000 t/a the registration deadline was 31 May 2013 [5]. For the tonnage bands 10 to 100 t/a and also 1 to 10 t/a the registration deadline will be 31 May 2018 [6]. A short overview on the registration deadlines for (late) pre-registered substances is given in Figure 4.7.

The relevant tonnage band for determination of the registration deadline is the highest amount of a certain substance manufactured of a European manufacture or imported by a single importer at least once after 1 June 2007 [7, 8].

Nota bene data requirements that correspond to the tonnage band can deviate from the relevant tonnage band for the registration deadline. If a substance was produced in the three consecutive years before the registration deadline, data requirements can be based on the average tonnage of the last three years before the year of registration [9]. When there had been a cessation and restart or an importer did not import the substance to be registered in three consecutive years before the registration deadline, then the estimated tonnage for the year of registration has to be considered.

This means there could be a case where a company had to register in 2010, but had only data requirements for the tonnage band 100 to 1000 t/a.

For example, when a company manufactured a certain substance for the first time in 2007 and relevant amounts per calendar year are for 2007: 990 t/a, 2008: 1000 t/a, 2009: 900 t/a, the registration deadline was 2010 because of the manufacture of 1000 t/a in 2008. Data requirements are based on the average tonnage of the years 2007 to 2009 and that is less than 1000 t/a, therefore data requirements in accordance with the tonnage band 100 to 1000 t/a have to be fulfilled.

The example shows that if the company would have manufactured in 2008 only 999 t/a instead of the 1000 t/a they would have had the registration deadline 31 May 2013. That would have meant that this company could have saved the money for preparation of the dossier and fees with ECHA for several years.

To watch carefully over the relevant amounts of a substance either manufactured by a European manufacturer or imported into the EU by an importer or even in the case of Non-European manufacturers that fulfill the registration obligations via an Only Representative it is highly recommended to do Substance Volume Tracking at least once per year.

There are several possibilities to cope with this issue. Bigger companies maybe will afford the implementation of highly effective IT-tools or amendments of already existing IT-tools, but in principle the task can also be fulfilled by having a simple table, as shown in Chapter 2, Figure 2.3.

4.4
Special Rules for Non-EU Manufacturers

As there are only REACH registration obligations for amounts of 1 t/a and more per manufacturer or importer, there are interesting options for globally acting enterprises. There is in several cases a big difference whether a substance is manufactured within the EU or outside of the EU, although in both cases the customers may be located within the EU. Figure 4.8 shows an example in which an EU manufacturer and a Non-EU manufacturer both manufacture the same amount of a certain substance and both of them sell half of the amount to customers located within the EU and the second part to customers outside of the EU.

	EU Manufacturer	Non-EU Manufacturer
Total tonnage manufactured within calendar year	100 t	100 t
Amount sold to EU customers	50 t	50 t
Amount sold to Non-EU customers	50 t	50 t
REACH relevance	EU manufacturer has to register 100 t	REACH relevance only concerning the amount that is delivered into the EU

Figure 4.8 REACH relevance for substances manufactured within the EU or outside of the EU.

	EU Manufacturer	Non-EU Manufacturer
Amount sold to EU customers	50 t	50 t
Number of EU customers	more than 50	more than 50
Amount that is bought from each customer	< 1 t/a	< 1 t/a
Registration obligation	EU Manufacturer has to register the total amount manufactured per calendar year	As every one of the EU customers can be considered as an importer under REACH, but every single customer sells less than 1 t/a, nobody has registration obligations

Figure 4.9 Example to illustrate.

Although the situation seems to be equal there arise different registration obligations under REACH depending on the relevant tonnage band for registration.

When a company located in France produces less than 1 t/a of a substance A, they do not have any registration obligation. When they manufacture 1 t/a or more they have to register the whole amount produced independent of where the customers are located.

If a company located in Switzerland manufactures 20 t/a of substance A and they sell part of this amount in Switzerland and other Non-EU countries and only a smaller part of this amount to EU customers, only the part of the substance that is delivered to EU customers has to be considered concerning REACH registration obligations. Either the Swiss company can register via an Only Representative within the EU or the customers in the role of the importer have to fulfill the registration obligations.

Figure 4.9 shows a further example concerning the different overall situation for manufacture of substances within the EU and outside of the EU although the distribution of the manufactured substance may be equal. If a company located in China produces 50 t/a of substance A and sells the whole 50 t/a to 51 customers

4 Obligation to Submit a Registration Dossier

within the EU on an equal share of the amount, every customer is an importer in the meaning of REACH, but each of them imports less than 1 t/a, therefore nobody will have any obligation to register substance A. Only when a customer in the role of the importer purchases 1 t/a or more of substance A will there occur REACH registration obligations. If a European importer imports substance A from different non-European manufacturers and buys less than 1 t/a from each supplier nevertheless it is possible that the importer will have REACH registration obligations as the whole amount of substance A that is imported per calendar year by this particular importer has to be considered (see also Figure 4.10).

If a European customer would like to purchase 2 t of a substance B (non-phase-in substance) in December 2011 from a non-European manufacturer who is not willing to register this substance via an Only Representative, in general the European customer in the role of the importer has to fulfill the registration obligations under REACH. In Figure 4.11 there is shown an alternative. If the European

	Importer 1	Importer 2
Amount of Substance A purchased from Non-EU Manufacturer 1	999 kg/a	999 kg/a
Amount of Substance A purchased from Non-EU Manufacturer 2	999 kg/a	-
Amount of substance A purchased from an EU manufacturer	50 t/a – but have not to be considered as Importer 1 in this case is not an Importer under REACH, but a Downstream User	-
Registration obligation	Registration obligation for importer 1, because the sum of imports is 1998 kg and therefore above 1 t/a. Importer 1 has to register substance A in the tonnage band 1 to 10 t/a	No registration obligation for Importer 2 as he imports less than 1 t/a

Figure 4.10 Amounts that have to be considered by importers.

	Importer 1	Importer 2
Amount of B imported in 2011	2 t	999 kg
Amount of B imported in 2012	-	999 kg
Amount of B used for a production campaign in January 2012	2 t	1998 kg
Registration obligation concerning B	yes	no

Figure 4.11 Import of 2 t of a Substance B.

customer purchases 999 kg of the substance B in December 2011 and another 999 kg in January 2012 there is no obligation to register at all. This means purchase in total 2 kg less and pay twice for the transportation of substance B, but there will be no registration obligations.

This could be interesting if the European customer could not manage to submit an Inquiry Dossier and also the registration dossier in due time in 2011, but has to start its production campaign in January 2012 with the purpose to deliver the product to his customer in time.

A further benefit could be that the European customer does not have to fear the publication of information from his registration dossier by ECHA. Confidentiality claims concerning confidential business information in a registration dossier would lead to higher fees with ECHA.

4.5 Consequences for Globally Acting Enterprises/What to Take into Account within a Decision-Making Process?

Very often it is with regard to economic aspects a great advantage to involve Non-European legal entities. Therefore, sometimes even a business place located in Switzerland can offer advantageous conditions, although production in Switzerland in general is assumed to be more expensive than in other countries of the world. A Swiss company is nearby the European customers but maybe registration obligations can be minimized by production in Switzerland. This will save time and money. Especially in the case of a multi-step synthesis a Non-European site has the great advantage that only the last step that is sold to an EU-customer needs to be considered concerning registration under REACH. All intermediate steps that are manufactured outside of the EU and are used for further processing outside of the EU do not have to be registered under REACH. The same can be applied to raw materials purchased from Non-EU countries, for example, from the Asian market.

For globally acting companies there will arise the question what to do on which site and when. By clever considerations the work-load and costs concerning REACH issues can be minimized.

Saving costs concerning REACH for the import of one or several raw materials, solvents and also for several intermediate steps in a multi-step synthesis can lead to a considerable cut down of the costs for a certain product.

4.6 Examples and Exercises

1. A company located in Germany manufactures 50 t/a of a substance A and imports a further 10 t/a of substance A from a manufacturer having its production site in France. Who has to register substance A and which amounts are relevant?

2. What changes when the German company mentioned in example 1 manufactures 60 t/a of substance A and purchases a further 10 t/a from a Chinese Manufacturer?

3. What has to be considered when a chemical company intends to manufacture a phase-in substance in 2014 for the first time?

4. A globally acting enterprise EXTRACLEVER has manufacturing sites in China, Germany, France and Switzerland. One of their customers is located within Italy and is interested in buying substance D. For the manufacturing of D EXTRACLEVER can purchase a raw material A from a company located in the United States and afterwards an intermediate B can be synthesized by using substance A. Substance B afterwards has to be transformed into the intermediate C that in the last step is transformed into Substance D.

What consequences may arise when EXTRACLEVER intends to produce Substance D at their site in China, Germany, France or Switzerland?

References

1 Regulation (EC) No 1907/2006 of the European Parliament and of the Council of 18 December 2006, concerning the Registration, Evaluation, Authorisation and Restriction of Chemicals (REACH), establishing a European Chemicals Agency, amending Directive 1999/45/EC and repealing Council Regulation (EEC) No 793/93 and Commission Regulation (EC) No 1488/94 as well as Council Directive 76/769/EEC and Commission Directives 91/155/EEC, 93/67/EEC, 93/105/EC and 2000/21/EC.
2 See Article 23 (1) (a) of the REACH regulation [1].
3 See Article 23 (1) (b) of the REACH regulation [1].
4 See Article 23 (1) (c) of the REACH regulation [1].
5 See Article 23 (2) of the REACH regulation [1].
6 See Article 23 (3) of the REACH regulation [1].
7 See Article 23 of the REACH regulation [1].
8 Guidance on registration, Version 2.0, May 2012, published by the European Chemicals Agency.
9 See Article 3 (30) of the REACH regulation [1].

5
Types of Registration

As under REACH [1] it is an obligation to cover all uses of the mentioned substance within the registration dossier it is important to have a good knowledge not only of the manufacturing process but also concerning all the steps in the lifecycle of the substance to be registered. When your company manufactures a substance that is used at your own site at a later stage it may be easy to fulfill the task listing all identified uses, whereas it is very often a time-consuming and difficult process for substances that are sold to a large number of customers who have a huge number of uses for this substance. If the manufacturer of a substance sells this substance at first to a trader or distributor in principle the trader/distributor should ask all his customers concerning their identified uses and afterwards report them in the other direction of the supply chain until the manufacturer/registrant to be receives the detailed information on how the substance is used. In reality, the process of asking customers that ask their customers and so on until the feedback will be given to the manufacturer can last several months and finally it could happen that the information passed on from the users to the supplier is not correct because the Downstream users are not experienced in reporting this sort of information in a way that could be used directly for the preparation of the dossier or maybe they made mistakes because they are not used to translating their uses into the Use Descriptor system [2] as it is foreseen within the REACH process. An example of a Letter from registrant to Downstream Users to ask for their identified uses can be found in Figure 5.1.

In some cases Downstream Users are competent and willing to list their identified uses, also they have not been asked for that by their suppliers. This could be clever especially when the Downstream User intends to choose only such suppliers that cover his uses by their registration. It is good practice to cover uses of Downstream Users in a registration, if the Downstream User informs the registrant in due time – at least 12 months before the registration deadline. After registration of a substance it is still possible to submit an Update to ECHA, if there are new identified uses that were not covered before, but it is time consuming and often costs in total are higher than for covering all identified uses in the first submission. An example of a Letter from a Downstream User to a supplier to make their uses identified uses/to ask whether the identified uses are covered can be found in Figure 5.2.

REACH Compliance – The Great Challenge for Globally Acting Enterprises, First Edition. Susanne Kamptmann.
© 2014 Wiley-VCH Verlag GmbH & Co. KGaA. Published 2014 by Wiley-VCH Verlag GmbH & Co. KGaA.

> **REACH – Questionnaire for Downstream Users to report their identified uses concerning [Substance], [CAS], [EINECS]**
>
> Dear customer,
>
> Our company has pre-registered [Substance] and now we start to prepare the registration dossier.
>
> As you know the REACH regulation demands that all uses of a substance are identified. For those substances that are subject to Chemical Safety Assessment it is necessary to develop Exposure Scenarios that are evaluated and that are communicated up and down the Supply Chain.
> In accordance with REACH Article 37 (2) it is the right of Downstream Users to make the uses of their substances known to their Suppliers in order to make them identified uses and have them considered in the registration dossier.
> Therefore, the next step for us in the ongoing registration process will be to collect the uses of our customers concerning the above stated substance.
>
> May we ask you to provide us with all your uses and if so also your customers´ uses? To enable quick editing of the requested information we attach a questionnaire that can be filled in easily.
>
> We would highly appreciate that you provide us with the requested information asap and preferably not later than [date]. Please fill in a separate questionnaire for each use of the substance.
>
> Thank you very much for your kind cooperation that will help us to register [substance] by considering your individual uses.
>
> Best regards,

Company			
Substance			
CAS		EINECS	
Substance is exempted from REACH registration obligations		☐ yes – please specify ☐ no – please go on with this questionnaire	
Use (Main user groups [3])		☐ **Industrial uses (SU 3)** ☐ **Professional uses (SU 22)** ☐ **Consumer uses (SU 21)**	
Sector of use (SU) [3]			
SU 1		Agriculture, forestry, fishery	
SU 2a		Mining, (without offshore industries)	
SU 2b		Offshore industries	
SU 4		Manufacture of food products	
SU 5		Manufacture of textiles, leather, fur	
SU 6a		Manufacture of wood and wood products	
SU 6b		Manufacture of pulp, paper and paper products	
SU 7		Printing and reproduction of recorded media	
SU 8		Manufacture of bulk, large-scale chemicals (including petroleum products)	

Figure 5.1 Letter from registrant to Downstream User concerning identified uses and template for the attached questionnaire.

SU 9		Manufacture of fine chemicals
SU 10		Formulation [mixing] of preparations and/or repackaging (excluding alloys)
SU 11		Manufacture of rubber products
SU 12		Manufacture of plastics products, including compounding and conversion
SU 13		Manufacture of other non-metallic mineral products, e.g. plasters, cement
SU 14		Manufacture of basic metals, including alloys
SU 15		Manufacture of fabricated metal products, except machinery and equipment
SU 16		Manufacture of computer, electronic and optical products, electrical equipment
SU 17		General manufacturing, e.g. machinery, equipment, vehicles, other transport equipment
SU 18		Manufacture of furniture
SU 19		Building and construction work
SU 20		Health services
SU 23		Electricity, steam, gas water supply and sewage treatment
SU 24		Scientific research and development
SU 0		Other → use NACE codes to specify (Link given in [3])

Product Category (PC) [4], [5]		
A) Chemical product category [4]		
PC1	Adhesives, sealants	
PC2	Adsorbents	
PC3	Air care products	
PC4	Anti-freeze and de-icing products	
PC7	Base metals and alloys	
PC8	Biocidal products (e.g. Disinfectants, pest control)	PC35 should be assigned to disinfectants being used as a component in a cleaning product
PC9a	Coatings and paints, thinners, paint removers	
PC9b	Fillers, putties, plasters, modeling clay	
PC9c	Finger paints	
PC11	Explosives	
PC12	Fertilizers	
PC13	Fuels	
PC14	Metal surface treatment products, including galvanic and electroplating products	This covers substances permanently binding with the metal surface
PC15	Non-metal-surface treatment products	Like for example treatment of walls before painting.
PC16	Heat transfer fluids	
PC17	Hydraulic fluids	
PC18	Ink and toners	
PC19	Intermediate	
PC20	Products such as pH-regulators, flocculants, precipitants, neutralization agents	This category covers processing aids used in the chemical industry
PC21	Laboratory chemicals	

Figure 5.1 (*Continued*)

PC23	Leather tanning, dye, finishing, impregnation and care products	
PC24	Lubricants, greases, release products	
PC25	Metal working fluids	
PC26	Paper and board dye, finishing and impregnation products: including bleaches and other processing aids	
PC27	Plant protection products	
PC28	Perfumes, fragrances	
PC29	Pharmaceuticals	
PC30	Photo-chemicals	
PC31	Polishes and wax blends	
PC32	Polymer preparations and compounds	
PC33	Semiconductors	
PC34	Textile dyes, finishing and impregnating products; including bleaches and other processing aids	
PC35	Washing and cleaning products (including solvent based products)	
PC36	Water softeners	
PC37	Water treatment chemicals	
PC38	Welding and soldering products (with flux coatings or flux cores), flux products	
PC39	Cosmetics, personal care products	
PC40	Extraction agents	
PC0	Other (use UCN codes: see Link in [4])	
	B) Product (Preparation Category) [5]	
	PC1: Adhesives, sealants	Glues, hobby use
		Glues DIY-use (carpet glue, tile glue, wood parquet glue)
		Glue from spray
		Sealants
	PC3: Air care products	Air care, instant action (aerosol sprays)
		Air care, continuous action (solid and liquid)
	PC9a: Coatings, paints, thinners, removers	Waterborne latex wall paint
		Solvent rich, high solid, waterborne paint
		Aerosol spray can
		Removers (paint-, glue-, wall paper-, sealant remover)
	PC9b: Fillers, putties, plasters, modeling clay	Fillers and putty
		Plasters and floor equalizers
		Modeling clay
	PC9c: Finger paints	Finger paints
	PC12: Fertilizers	Lawn and garden preparations
	PC13: Fuels	Liquids
	PC24: Lubricants, greases, release products	Liquids
		Pastes
		Sprays
	PC31: Polishes and wax blends	Polishes, wax/cream (floor, furniture, shoes)
		Polishes, spray (furniture, shoes)

Figure 5.1 *(Continued)*

PC35: Washing and cleaning products (including solvent based products)	Laundry and dish washing products	
	Cleaners, liquids (all purpose cleaners, sanitary products, floor cleaners, glass-cleaners, carpet cleaners, metal cleaners)	
	Cleaners, trigger sprays (all-purpose cleaners, sanitary products, glass cleaners)	
Process category (PROC) [6]		
PROC1	Use in closed process, no likelihood of exposure	Use of the substances in high integrity contained system where little potential exists for exposures, e.g. any sampling via closed loop systems
PROC2	Use in closed, continuous process with occasional controlled exposure	Continuous process but where the design philosophy is not specifically aimed at minimizing emissions It is not high integrity and occasional expose will arise, e.g. through maintenance, sampling and equipment breakages
PROC3	Use in closed batch process (synthesis or formulation)	Batch manufacture of a chemical or formulation where the predominant handling is in a contained manner, e.g. through enclosed transfers, but where some opportunity for contact with chemicals occurs, e.g. through sampling
PROC4	Use in batch and other process (synthesis) where opportunity for exposure arises	Use in batch manufacture of a chemical where significant opportunity for exposure arises, e.g. during charging, sampling or discharge of material, and when the nature of the design is likely to result in exposure
PROC5	Mixing or blending in batch processes for formulation of preparations and articles (multistage and/or significant contact)	Manufacture or formulation of chemical products or articles using technologies related to mixing and blending of solid or liquid materials, and where the process is in stages and provides the opportunity for significant contact at any stage
PROC6	Calendering operations	Processing of product matrix Calendering at elevated temperature and large exposed surface
PROC7	Industrial spraying	Air-dispersive techniques Spraying for surface coating, adhesives, polishes/cleaners, air-care products, sandblasting Substances can be inhaled as aerosols. The energy of the aerosol particles may require advanced exposure controls; in case of coating, overspray may lead to waste water and waste.
PROC8a	Transfer of substance or preparation (charging/discharging) from/to vessels/large containers at non-dedicated facilities	Sampling, loading, filling, transfer, dumping, bagging in non-dedicated facilities. Exposure related to dust, vapour, aerosols or spillage, and cleaning of equipment to be expected.

Figure 5.1 (*Continued*)

PROC8b	Transfer of substance or preparation (charging/discharging) from/to vessels/large containers at dedicated facilities	Sampling, loading, filling, transfer, dumping, bagging in dedicated facilities. Exposure related to dust, vapour, aerosols or spillage, and cleaning of equipment to be expected.
PROC9	Transfer of substance or preparation into small containers (dedicated filling line, including weighing)	Filling lines specifically designed to both capture vapour and aerosol emissions and minimize spillage
PROC10	Roller application or brushing	Low-energy spreading of e.g. coatings Including cleaning of surfaces. Substances can be inhaled as vapors, skin contact can occur through droplets, splashes, working with wipes and handling of treated surfaces.
PROC11	Non industrial spraying	Air dispersive techniques Spraying for surface coating, adhesives, polishes/cleaners, air care products, sandblasting Substances can be inhaled as aerosols. The energy of the aerosols particles may require advanced exposure controls.
PROC12	Use of blowing agents in manufacture of foam	
PROC13	Treatment of articles by dipping and pouring	Immersion operations Treatment of articles by dipping, pouring, immersing, soaking, washing out or washing in substances; including cold formation or resin type matrix. Includes handling of treated objects (e.g. after drying, plating). Substance is applied to a surface by low energy techniques such as dipping the article into a bath or pouring a preparation onto a surface.
PROC14	Production of preparations or articles by tabletting, compression, extrusion, peletization	Processing of preparations and/or substances (liquid and solid) into preparations or articles. Substances in the chemical matrix may be exposed to elevated mechanical and/or thermal energy conditions. Exposure is predominantly related to volatiles and/or generated fumes, dust may be formed as well.
PROC15	Use as laboratory agent	Use of substances at small scale laboratory (<1l or 1 kg present at workplace). Larger laboratories and R+D installations should be treated as industrial processes.
PROC16	Using material as fuel sources, limited exposure to unburned product to be expected	Covers the use of material as fuel sources (including additives) where limited exposure to the product in its unburned form is expected. Does not cover exposure as a consequence of spillage or combustion.

Figure 5.1 (*Continued*)

PROC17	Lubrication under high-energy conditions and in partly open process	Lubrication at high-energy conditions (temperature, friction) between moving parts and substance; significant part of process is open to workers. The metal-working fluid may form aerosols or fumes due to rapidly moving metal parts.
PROC18	Greasing at high energy conditions	Use as lubricant where significant energy or temperature is applied between the substance and the moving parts
PROC19	Hand-mixing with intimate contact and only PPE available	Address occupations where intimate and intentional contact with substances occurs without any specific exposure controls other than PPE.
PROC20	Heat and pressure transfer fluids in dispersive, professional use but closed systems	Motor and engine oils, brake fluids Also in these applications, the lubricant may be exposed to high energy conditions and chemical reactions may take place during use. Exhausted fluids need to be disposed of as waste. Repair and maintenance may lead to skin contact.
PROC21	Low energy manipulation of substances bound in materials and/or articles	Manual cutting, cold rolling or assembly/disassembly of material/article (including metals in massive form), possibly resulting in the release of fibres, metal fumes or dust
PROC22	Potentially closed processing operations with minerals/metals at elevated temperature Industrial setting	Activities at smelters, furnaces, refineries, coke ovens. Exposure related to dust and fumes to be expected. Emission from direct cooling may be relevant.
PROC23	Open processing and transfer operations with minerals/metals at elevated temperature	Sand and die casting, tapping and casting melted solids, drossing of melted solids, hot dip galvanizing, raking of melted solids in paving Exposure related to dust and fumes to be expected.
PROC24	High (mechanical) energy work-up of substances bound in materials and/or articles	Substantial thermal or kinetic energy applied to substance (including metals in massive form) by hot rolling/forming, grinding, mechanical cutting, drilling or sanding. Exposure is predominantly expected to be to dust. Dust or aerosol emission as result of direct cooling may be expected.
PROC25	Other hot work operations with metals	Welding, soldering, gouging, brazing, flame cutting Exposure is predominantly expected to fumes and gases.
PROC26	Handling of solid inorganic substances at ambient temperature	Transfer and handling of ores, concentrates, raw metal oxides and scrap; packaging, un-packaging, mixing/blending and weighing of metal powders or other minerals
PROC27a	Production of metal powders (hot processes)	Production of metal powders by hot metallurgical processes (atomization, dry dispersion)

Figure 5.1 (*Continued*)

PROC27b	Production of metal powders (wet processes)	Production of metal powders by wet metallurgical processes (electrolysis, wet dispersion)

Article category (AC)[7], [8]	
A) For substances in articles with no intended release [7]	
1) Categories of complex articles	
AC1	Vehicles
AC2	Machinery, mechanical appliances, electrical/electronic articles
AC3	Electrical batteries and accumulators
2) Categories of material based articles	
AC4	Stone, plaster, cement, glass and ceramic articles
AC5	Fabrics, textiles and apparel
AC6	Leather articles
AC7	Metal articles
AC8	Paper articles
AC10	Rubber articles
AC11	Wood articles
AC13	Plastic articles
Other	Specify (see Link in [7])
B) For substances in articles with intended release [8]	
AC30	Other articles with intended release of substances, please specify
AC31	Scented clothes
AC32	Scented eraser
AC34	Scented toys
AC35	Scented paper articles
AC36	Scented CD
AC38	Packaging material for metal parts, releasing grease/corrosion inhibitors

Temperature of the process		°C

Duration and frequency of exposure	
	☐ 5 days per week
	☐ 4 hours/day
	☐ 8 hours/day
	☐ other (specify):

Exposure to the substance in preparations or articles	
Concentration of the substance	
Physical state	
Applied amount of the substance (per application, per time or per activity)	
Indoor or outdoor use and short description	

Risk management measures for human health	
Technical measures	☐ Open process ☐ Closed process ☐ Automated process ☐ general ventilation ☐ Local exhaust ventilation ☐ Other (specify):
Organizational measures	☐ Time of operations/activities limited to [time] ☐ Other (specify):

Figure 5.1 (Continued)

Personal protection measures	☐ Gas filter masks
	☐ Dust filter masks
	☐ Goggles
	☐ Gloves
	☐ Protective clothing (specify):
	☐ Other (specify):
Consumer related measures	☐ Form of packaging (specify):
	☐ Migration-preventing coating (specify):
	☐ Other (specify):

Risk management measures for effluents and waste	
A) For liquid waste	
Emission to a sewage treatment plant	☐ yes
	☐ no
	Quantity: l/day
	Concentration
	Duration of emission: days/year
Type of sewage treatment	☐ physico-chemical treatment
	☐ biological treatment
	☐ Other (specify):
B) For solid waste or gaseous waste	
Type of treatment (short description)	
	Quantity: l/day
	Concentration
	Duration of emission: days/year

Figure 5.1 (*Continued*)

Dear Supplier,

We are purchasing [substance], [CAS], [EINECS] from your company.
In accordance with REACH Article 37 (2) it is the right of Downstream Users to make the uses of their substances known to their Suppliers in order to make them identified uses and have them considered in the registration dossier.
We use the above-mentioned substance
☐ as a transported isolated intermediate in accordance with REACH Article 18 (4).
   Amount: t/a
☐ as a solvent, a monomer for the manufacturing of polymers, a catalyst.
      This use can be described by the following Use Descriptors [SU], [PROC], [PC]…
   Amount: t/a
☐ outside of the scope of REACH. [Justification]. Amount: t/a

Please confirm in writing that our uses will be covered with your registration.

Best regards,

Figure 5.2 Letter from Downstream User to Supplier to make their uses identified uses.

However, depending on the uses to be covered the registrant has to make a decision which type of registration will be appropriate to cover the whole lifecycle from manufacturing until use of the substance.

5.1
Standard Registration, Full Registration or Registration as a Substance

A so called "standard registration" or "full registration", also known as "registration as a substance", is the general type of registration. All identified uses can be covered.

If a substance is not used for the synthesis of another substance in any case a full registration will be necessary. This is the case for all substances used as a solvent. The same applies for substances used as catalysts as they are not intentionally transformed into another substance.

Polymers themselves are exempted from REACH, but instead the monomers that were used to manufacture the polymer have to be registered under REACH and there is the general obligation to do a full registration for these monomers.

All substances that have widespread uses and are handled by consumers have to be registered as a substance. This sort of handling is defined as an end use.

Substances contained in articles with intended release also fall under this type of registration.

The identified uses that are covered with the registration will be listed in the SDS after registration. For a full registration the registrant provides an Annex to the well-known SDS. In this annex the identified uses and exposure scenarios are listed. This often bulky document is called an extended Safety Data Sheet (eSDS).

Data requirements for the preparation of the registration dossier are dependent on the tonnage band. Fees with ECHA are differentiated corresponding to the tonnage band. Detailed information on data requirements for this type of registration will be provided in Chapter 6 of this book.

5.2
Registration as an On-site Isolated Intermediate

Manufacturers located within the EU have the obligation to register intermediates that are isolated and afterwards are used on the same site, although they never will be marketed.

If the registrant can prove that he always abides strictly controlled conditions and acts in accordance with REACH Article 17 (3) [9] he is allowed to benefit from reduced data requirements for his registration compared to the standard registration.

On the contrary, a manufacturer located in Switzerland is allowed to do multi-step reactions with isolation of all intermediate steps without having any obligation to register these substances.

The European manufacturer having registration obligations at the same time has to face the fact that some of the data submitted with the registration dossier will be disseminated and already before submission of the dossier to ECHA other (pre-) registrants of the same substance will know their competitors really well, whereas Non-European manufacturers can act in secret concerning their intermediates.

5.3
Registration as a Transported Isolated Intermediate

As there are reduced data requirements and a registrant can also benefit from a reduced fee for the registration of a transported isolated intermediate compared to a full registration, it is a topic of high interest for industry. If a registrant wants to benefit from this inexpensive type of registration several conditions have to be fulfilled.

First of all such a substance must be used completely for the synthesis of another substance. That means it can be deemed to be an intermediate. If this can be confirmed the registrant has two further obligations – he has to confirm that the substance during its manufacturing process is used under strictly controlled conditions in accordance with Article 18(4) of the REACH regulation [10] and he has to confirm this for the whole lifecycle of the substance. When the substance is purchased by customers they have to confirm the handling in accordance with Article 18(4) [10] for the rest of the supply chain. This means when a trader/distributor buys a transported isolated intermediate from a supplier he has to ask all his customers for their confirmation and store them carefully and he has also to pass on a confirmation to his supplier. Every actor in the supply chain is responsible to abide by the conditions he has confirmed. The supplier is not obliged to control the correctness of the confirmations he received from his customers by an on-site inspection at the customer's site, it is enough to demand the confirmation itself from the Downstream User (next actor in the supply chain) and to pass it on to national authorities on request. It is here also very important to respect the competition law. If a manufacturer sells a transported isolated intermediate to a trader or distributor it will clearly not be in accordance with the competition law to ask for the confirmation of the customers that are purchasing from this trader or distributor.

An example of a letter that can be sent to your customers asking them for confirmation of using a certain substance as a transported isolated intermediate under strictly controlled conditions can be found in Figure 5.3.

As REACH is valid only within the EU (including the EEA) in reality it could be difficult to obtain such a confirmation from Non-European customers. But practice showed that Non-European customers are often well aware of REACH matters and are willing to sign the requested confirmation, because they will directly benefit from lower costs for registration obligations of their suppliers, as the purchased substance will be less expensive. Non-EU customers can also be relaxed as

> **REACH – Request for Confirmation from Downstream User to Supplier concerning substance [substance name], CAS [CAS number], EC [EINECS number]**
>
> Dear customer,
>
> Our company has registered/intends to register the above mentioned substance as a transported isolated intermediate to benefit from reduced data requirements for the registration. Furthermore, intermediates will not be subject to any authorization process.
> According to Article 18 (4) of the REACH regulation this type of registration can only be applied, if we, as manufacturer, obtain from all our customers a confirmation that the synthesis of (an)other substance(s) from this intermediate takes place on your site(s) under strictly controlled conditions as described in Article 18 (4) of the REACH regulation.
>
> **The substance will not be registered for any other use.**
>
> To ensure uninterrupted supply of the above mentioned substance we kindly ask you for your written confirmation that the substance [substance name] delivered by our company is solely used as a transported isolated intermediate at your site and compliance with the strictly controlled conditions is given at your site.
>
> Please complete the attached confirmation and return it asap.
>
> Thank you very much for your cooperation that will help us to deliver the desired product to your company without undue delay.
>
> Best regards,

Figure 5.3 Letter to ask for Confirmation concerning Article 18 (4) and template for the confirmation.

they cannot be controlled by ECHA or other European authorities. Therefore, even if they would not abide by strictly controlled conditions as demanded by REACH Article 18 (4) nobody would be in the position to go to law for this. A registrant having the mentioned confirmation from all his customers will without any doubt be in a good position. Problems could arise in cases where customers are not willing to sign a confirmation concerning REACH Article 18 (4). In case the Non-EU customers do not provide their European suppliers with an Article 18 (4) confirmation, the registrant may send a letter to these customers to state clearly that it is expected that the substance is handled accordingly. For an example letter see Figure 5.4. Hopefully national authorities will accept this as this problem is not solved satisfactory for registrants manufacturing a substance within the EU. Whereas a European manufacturer is obliged to register the whole amount that is manufactured at his site and also has to provide a proof for that in case of on-site inspections, a Non-EU manufacturer who registers via an Only Representative has to consider only the part of the substance that is brought into the EU later, but not parts of the substance that are manufactured at his site and afterwards are directly sold to Non-EU customers.

Please return to:
[Supplier and address]

Confirmation concerning Article 18 (4) of the REACH regulation concerning the substance [substance name], CAS [CAS number], EC [EINECS number]

Customer/downstream User details
Company name:
Address:
ZIP Code/City:

☐ Use of the above mentioned substance as a transported isolated intermediate: Amount (w/w%): _____ %
☐ Use that requires a standard registration: Amount (w/w%): _____ %
  → Our company will take care of the CSR on our own
☐ Use of the above mentioned substance out of the scope of REACH: Amount (w/w%): _____ %

[x] Herewith we confirm that the substance [substance name], CAS [CAS number], EC [EINECS number] is used by our company and/or our customers in the proportion indicated above exclusively for the synthesis of (an)other substance(s) under strictly controlled conditions as defined in Article 18 (4) of Regulation (EC) No. 1907/2006 (REACH regulation):
  a) The substance is rigorously contained by technical means during its whole lifecycle including manufacture*), purification, cleaning and maintenance of equipment, sampling, analysis, loading and unloading of equipment or vessels, waste disposal or purification and storage.
  b) Procedural and control technologies shall be used that minimize emission and any resulting exposure.
  c) Only properly trained and authorized personnel handle the substance.
  d) In the case of cleaning and maintenance works, special procedures such as purging and washing are applied before the system is opened and entered.
  e) In cases of accident and where waste is generated, procedural and/or control technologies are used to minimize emissions and the resulting exposure during purification or cleaning and maintenance procedures.
  f) Substance-handling procedures are well documented and strictly supervised by the site operator.

We are aware that internal documentation is necessary to fulfil our obligations as a Downstream User and that this documentation has to be provided to the competent authorities on request.

[ ] Herewith we confirm that the substance [substance name], CAS [CAS number], EC [EINECS number] in the proportion indicated above supplied by your company that is not used as a transported isolated intermediate, is solely used under the following exemption from REACH:
[justification]

This document shall be valid from the date of signature without any expiration/until we notify you of any change.

[Date and place]
Signed: [signature]
By: [name]
Title: [title]

*) "manufacture" may be omitted as this is not applicable to describe the use of the substance by a Downstream User.

Figure 5.3 (*Continued*)

> Der Customer,
>
> Our company is located within the EU and therefore we have the obligation to register under the European Chemical legislation REACH all substances that are manufactured at our site.
> We have registered the substance [substance name], CAS [CAS number], EINECS/EC [EINECS number] as a transported isolated intermediate in accordance with REACH Article 18 (4).
> The registration of this intermediate is subject to its use under Strictly Controlled Conditions according to Article 18 (4) of the REACH regulation (Regulation (EC) No 1907/2006).
>
> **Please ensure that this substance is only used as an intermediate for the synthesis of another substance under Strictly Controlled Conditions.**
>
> Best regards,

Figure 5.4 Letter that may be sent to Non-EU customers that are not willing to support their EU supplier by signing an Article 18 (4) confirmation.

It is also necessary that a manufacturer has on-site documentation concerning the manufacturing process to demonstrate that the substance is manufactured in accordance with Article 18 (4) of the REACH regulation [10] and that a proof of that is in the registration dossier. In the past it was enough to give a short statement in the registration dossier that the substance is manufactured and handled under strictly controlled conditions in the meaning of REACH Article 18 (4). However, in the meantime ECHA recommends to have a separate document with more detailed information that could be attached to the registration dossier. An example of such documentation can be found in the Guidance on intermediates [11].

Whereas in Appendix 1 [12] of the mentioned Guidance document you can find a list of issues that maybe taken into consideration for checking that the isolated intermediates are manufactured and used under strictly controlled conditions, Appendix 2 gives an example of inhouse documentation [12]. An overview on Appendix 1 can be found in Figure 5.5, whereas a format for the inhouse documentation is presented in Figure 5.6. This inhouse documentation is an appropriate possibility for manufacturers/importers to prove that they are allowed to benefit from reduced data requirements for the registration of the mentioned substance as a transported isolated intermediate. This format can also be used for on-site isolated intermediates. It is highly recommended to have this documentation available in case of on-site inspections by national authorities.

Finally, in Appendix 3 [14] of the Guidance on intermediates there is an example of documenting information on risk management in a registration dossier for isolated on-site and transported intermediates. Companies can feel free to use this format directly from the Guidance on intermediates for their purposes or they can adapt them slightly to have the information clearly arranged. For companies that in general have large numbers of substances that are manufactured or imported

Substance name (internal abbreviation)
Trade name/synonyms   structural formula
CAS [ ]
EINECS [ ]
Molecular formula

1. **Has the lifecycle of the substance been accounted for?**

   • Manufacture of the intermediate?
   Continuous process or batch operation? Scale of operation? [12]

   • Use of the intermediate?
   Continuous process or batch operation? Scale of operation? [12]

   • Final synthesis process? [12]

   • Any purification step? [12]

   • Sampling and analysis? [12]

   • Loading and unloading from equipment or vessels and any other substance transfers? [12]

   • Any relevant storage? [12]

   • Waste treatment? [12]

2. **Is rigorous containment by technical means in place?**

   • The substance is rigorously contained by the following means [12]

   • Procedures to ensure containment have been applied and maintained for all stages of production and processing [12]

   • Management system is in place [12]

   • Implementation of existing EU legislation [12]

   • Monitoring measurements to check on potential remaining emissions are being carried out.
   This includes: [12]

3. **Are procedural and control technologies being used to minimize emissions?**
   • Residual emissions from rigorous containment occur at the following steps of the processes.....[12]
   These emissions are minimized by the following procedural and control techniques (differentiation regarding workplaces and environment required): [12]
   • Emissions from purification, cleaning and maintenance after accidents are minimized by the following procedural and control techniques (differentiation regarding workplaces and environment required): .....[12]
   • Emissions from purification, cleaning and maintenance are minimized by the following procedural and control techniques (differentiation regarding workplaces and environment required): .....[12]
   • Emissions from waste handling are minimized by the following procedural and control techniques (differentiation regarding workplaces and environment required): [12]

4. **Are only properly trained and authorized personnel handling the substance?**
   • Relevant training or authorization scheme covers this substance and/or process [12]

Figure 5.5 List of issues that may be taken into consideration for checking that an isolated intermediate is manufactured and used under strictly controlled conditions [15].

- A procedure ensures that only trained and authorized persons handle the substance [12]
- Other legislative frameworks that control the handling of the substance have been considered [12]

5. **Are special procedures applied before the system is opened and entered during cleaning and maintenance works?**
   - Process procedures for containment during cleaning and maintenance have been accounted for in plant and engineering design as appropriate for the site
   - Operational procedure system checks include cleaning and maintenance of process equipment
   - Risk-management measures are applied during cleaning and maintenance
   - Specific procedures before the system is opened. These include, e.g. purging and washing and .... (further specify)

6. **Are substance-handling procedures well documented and supervised by the site operator?**
   - Occupational procedures have been assessed and are documented

7. **For transported isolated intermediates:**
   - Confirmation that the synthesis of (an)other substance(s) from which intermediate takes place under strictly controlled conditions on other sites has been documented

Figure 5.5 (Continued)

it is helpful to complete the format with an own header containing information on the substance (substance name, internal abbreviation, trade name, synonyms, CAS number, EC number, molecular formula, structural formula and maybe statements concerning the role of the registrant: acts as manufacturer or importer or both, etc.) to be considered. Therefore, you will find in Figure 5.7 an example of that type of documentation based on Appendix 3 in the Guidance on intermediates [16], but completed with a helpful header.

As ECHA is interested to have proof of the manufacturing process in accordance with the conditions given in Article 18 (4) of the REACH regulation [10] it is highly recommended to prepare a document as shown in Appendix 3 of the Guidance on intermediates [14] (see also Figure 5.7) that may be attached in Chapter 13 of the IUCLID substance data set instead of the Chemical Safety Report that is necessary for full registrations.

The confirmations that a manufacturer receives by the customers or Downstream Users have not to be included in the registration dossier, but they must be stored by the registrant and on request they have to be shown to national authorities.

So far, we considered the situation for a registrant that manufactures a transported isolated intermediate within the EU. The situation is slightly different for a non-European manufacturer that intends to register via an Only Representative within the EU. First, there has to be taken into account only the amount of the substance that is brought into the EU – only this amount has to be registered and for the registration as a transported isolated intermediate only for this part of the substance manufactured outside the EU the registrant needs the confirmations concerning Article 18(4) [10] from the Downstream user(s). In practice, there can

Substance name (internal abbreviation)
Trade name/synonyms   **structural formula**
CAS [ ]
EINECS [ ]
Molecular formula

1. **Description of technological process used in manufacture**

2. **Description of the uses of the substance**

    • Give a description of the uses of the substance on the different sites. [13]
    Check that any relevant storage, processing and the synthesis process of the final substance have been accounted for. [13]

3. ***Is the substance rigorously contained:***
    • During the manufacturing process? [13]
       • Description of the process and technical means to contain the substance [13]
       • Identification of potential emissions to: [13]
          • Workplace [13]
          • Environment [13]
       • Modeling estimations or available monitoring data if needed [13]
       • Procedure and systems in place to comply with existing health, safety and environmental legislation. [13]
    • During the use? [13]
       • Description of the process and technical means to contain the substance. [13]
    • Identification of potential emission to: [13]
       • Workplace [13]
       • Environment (air, waste, water, soil, etc.) [13]

    • Modeling estimations or available monitoring data if needed [13]

    • During substance transfers before and after transport? [13]
       • Description of the process and technical means to contain the substance. [13]
    • Identification of potential emissions to: [13]
       • Workplace [13]
       • Environment (air, wastewater, soil, etc.) [13]
    • Modeling estimations or available monitoring data if needed. [13]

4. **If emissions have been identified on sites of manufacture or uses, are there procedural and control technologies to minimize emission and resulting exposure?**
    • Give a description of these procedural and control technologies in place, including those applied after accidents and for wastewater collection and treatment. [13]

5. **Is the substance handled by trained and authorized personnel?**

    • Are the personnel provided with safety data sheet (SDS) of the substances handled? [13]
    • Is there sufficient training and information on appropriate precautions and working procedures (proper labeling of specific working places) at workplace? [13]

    • Is it guaranteed that only trained personnel handle dangerous substances? [13]

    Give a description of the information and training in place. [13]

Figure 5.6 Example format recommended to registrants for documenting inhouse information on strictly controlled conditions [17].

Substance name (internal abbreviation)
Trade name/synonyms   structural formula
CAS [ ]
EINECS [ ]
Molecular formula

1. **Brief description of technological process applied in manufacture of the intermediate**
   - Provide an overall technical description (no details). A simple overview scheme may support understanding. Ensure that all relevant activities (unit operations) are covered in this description, such as synthesis, purification steps, cleaning and maintenance, sampling and analysis, loading and unloading, storage and waste treatment [14]

2. **Brief description of technological processes applied in use of the intermediate**
   - Provide an overall technical description (no details). A simple overview scheme may support understanding. Ensure that all relevant activities (unit operations) are covered in this description, such as synthesis, purification steps, cleaning and maintenance, sampling and analysis, loading and unloading, storage and waste treatment [14]

3. **Means of rigorous containment and minimization technologies applied by the registrant during the manufacturing and/or use process**
   - Description of the technical means to rigorously contain the substance. Make reference to different activities (unit operations) and lifecycle stages as appropriate
   - Identification of residual emissions to:
     - Workplace
     - Environment (air, onsite water streams)
   - Description of the procedural and control technologies in place to minimize emission and resulting exposure. A rough quantification of the releases and information on effectiveness of control techniques may be useful to demonstrate that the technologies ensure rigorous containment and minimization of releases.
     - Workplace
     - Environment (air, waste water, discharge from site)
   - Specify the management means and training that particularly contribute to the functioning of the technical means described above. [14]

4. **Means of rigorous containment and minimization technologies recommended to the user of the intermediate:**
   - Description of the technical means to rigorously contain the substance. Make reference to the different lifecycle stages and activities (unit operations) as appropriate
   - Identification of residual emissions to:
     - Workplace
     - Environment (air, onsite water streams)
   - Description of the procedural and control technologies in place to minimize emission and resulting exposure? A rough quantification of the releases and information on effectiveness of control techniques may be useful to demonstrate that the technologies ensure rigorous containment and minimization of releases
     - Workplace
     - Environment (air, waste water discharge from site)
   - Specify the management means and training that particularly contribute to the functioning of the technical means described above.
   - Are these or other procedures communicated to the user of the intermediates? [14]

5. **Special procedures applied before cleaning and maintenance**
   - Description of the special procedures (such as purging and washing) applied before the system (any contained operation units within the lifecycle of the

Figure 5.7 Example format recommended to registrants for documention information on risk management (to be attached to a registration dossier for isolated on-site and transported intermediates) [16].

substance) is opened and entered for cleaning and maintenance work.
- Are these or other procedures communicated to the user of the intermediates? [14]

6. Describe activity and type of PPE in case of accidents, incidents, maintenance and cleaning activities
- Briefly list the activities and required type of PPE for the situations mentioned above (no details required).
- Are these or other procedures and suitable PPE communicated to the user of the intermediates? [14]

7. Waste information
- Identify the process stages where waste is generated (e.g. purification, maintenance, emission controls). Briefly describe the type of treatment applied onsite.
- Briefly describe the type of treatment applied offsite.
- A rough quantification of waste amounts may be useful to demonstrate that the technologies ensure rigorous containment and minimization of releases.

Figure 5.7 (Continued)

occur difficulties, if the Non-EU manufacturer sells such a substance also to Non-EU customers and those customers afterwards sell the substance to their customers and perhaps they are located within the EU. It is recommended that the Non-EU manufacturer gives the registration number (normally in the SDS) only to the European customers and not to Non-European customers, especially when the registrant has chosen a smaller tonnage band than the highest one, but manufactures in total more than the registered tonnage band. Otherwise, there could arise situations in which a Non-European trader sells a substance to an EU customer pretending that it is already registered although the amount was not covered by the registration done from the Non-EU manufacturer via his Only Representative. A special case could also be a Non-EU manufacturer who has several customers within the EU. When one of the customers registers the substance on its own (therefore acts as an importer under REACH) and all other customers would rather be Downstream users, the Non-EU manufacturer is allowed to register via his Only Representative just that part of the substance that is manufactured for the customers that are not registered themselves. In practice, this could lead to the situation where one/or several of the EU customers and also the Non-EU manufacturer via Only Representative can each legally register in a smaller tonnage band than it would be necessary for a manufacturer located within the EU producing the same amount per year as the Non-EU manufacturer does. For a transported isolated intermediate this could be an advantage when the amount manufactured per calendar year is 1000 t/a and more, but each registrant can register in the tonnage band below 1000 t/a, because data requirements then are reduced compared to the higher tonnage band. However, the financial benefit from such a construct mostly will not be decisive for a transported isolated intermediate – but the same strategy will work for full registrations and then result in a considerable advantage.

5.4
Formerly Notified Substances

Formerly notified substances (Directive 67/548/EEC) can be regarded as registered under REACH – registrant needs only to claim a registration number with ECHA. Then the substance is registered under REACH either in the tonnage band mentioned in the former notification or at least in the smallest tonnage band of 1 to 10 t/a as a substance. This means it is a standard registration at least until there is made a further update after the claim of registration number.

5.5
PPORD

PPORD is the abbreviation for Product and process orientated research and development [18]. A PPORD notification is not really a type of registration, but for the sake of completeness it shall be mentioned here.

A substance manufactured or used solely for this purpose may be exempted from registration obligations for a certain time when there is submitted a PPORD notification to ECHA. The great advantage of a PPORD notification compared to a regular registration is that ECHA will not disseminate information submitted to them within the PPORD notification. Confidential business information can be kept secret at least for the time of validity of the PPORD notification. For the preparation of a PPORD notification less data, less information and less time are necessary than for the preparation of a registration dossier, especially for non-phase-in substances. Fees with ECHA are low for this type of submission. The only disadvantage may be that at a later stage, nevertheless, a registration has to be done, but a company will then be in the position to know whether the substance can be marketed and whether they will be in the position to earn enough money with this product to be in the position to afford the costs connected to the registration obligations.

When the amount of a substance is handled within the scope of scientific research and development in volumes of less than 1 t/a [19] there is no need to submit a notification to ECHA.

Further details concerning PPORD notifications and data requirements can be found in Chapter 6 of this book.

5.6
Examples and Exercises

1. Your company intends to register a substance that is used as a catalyst at your own site in amounts of >1 t/a. Which type of registration may be appropriate?

2. A European company manufactures benzoic acid chloride that afterwards is used by their customers as an intermediate for the manufacturing of a pharmaceutical intermediate. Which type of registration may be appropriate? Are there special conditions to be fulfilled? Please give a justification for your decision.

3. Are there any registration obligations when a manufacturer produces a substance that afterwards is completely used at his own site? Why?

4. Which types of registration do you know? For which of these types may a Non-EU manufacturer be in the position to appoint an Only Representative that will register on behalf of the Non-EU manufacturer?

5. For which types of registration can an Only Representative do a registration under REACH on behalf of a Chinese trader?

6. Within the group of a professor at the University of Groningen PhD students produce a lot of new substances every year. Some of them are used for the synthesis of other substances, others will be tested concerning their biological activity at the site of pharmaceutical companies. What substances need to be registered by the professor and which type of registration will be appropriate? Why?

7. A manufacturer M manufactures 105 t/a of a substance E. 95 t/a are used as a solvent, 7 t/a are used for synthesis of another substance, 3 t/a are used under the scope of medicinal products or in food and foodstuffs. How can M achieve REACH compliance.

8. It is intended to do a multi-step synthesis within a company group to manufacture finally a substance D. First, a raw material A shall be imported from India. Substance A is transferred into substance B and afterwards B shall be transported to another site; at the other site substance B shall be used for the synthesis of substance C. C afterwards is either isolated or not isolated, but will in any case be used for the synthesis of substance D.

D is sold and transported to a customer. The customer may use substance D as an intermediate or as a catalyst.

Assume that the company group that manufactures substance D has all its production sites in Europe and the customer D is also located in Europe.

Which type of registration is then appropriate for Substances A, B, C and D? Give a short justification for your decision.

References

1 Regulation (EC) No 1907/2006 of the European Parliament and of the Council of 18 December 2006, concerning the Registration, Evaluation, Authorisation and Restriction of Chemicals (REACH), establishing a European Chemicals Agency, amending Directive 1999/45/EC and repealing Council Regulation (EEC)

No 793/93 and Commission Regulation (EC) No 1488/94 as well as Council Directive 76/769/EEC and Commission Directives 91/155/EEC, 93/67/EEC, 93/105/EC and 2000/21/EC.

2 Guidance on information requirements and chemical safety assessment Chapter R.12: Use descriptor system, Version: 2, March 2010, published by the European Chemicals Agency.
3 See Appendix R.12-1 of the Guidance on information requirements [2].
4 See Appendix R.12-2.1 of the Guidance on information requirements [2].
5 See Appendix R.12-2.2 of the Guidance on information requirements [2].
6 See Appendix R.12-3 of the Guidance on information requirements [2].
7 See Appendix R.12-5.1 of the Guidance on information requirements [2].
8 See Appendix R.12-5.2 of the Guidance on information requirements [2].
9 See Article 17 (3) of the REACH regulation [1].
10 See Article 18 (4) of the REACH regulation [1].
11 Guidance on intermediates, Version:2, December 2010, published by the European Chemicals Agency.
12 See Appendix 1 of the Guidance on intermediates [11].
13 See Appendix 2 of the Guidance on intermediates [11].
14 See Appendix 3 of the Guidance on intermediates [11].
15 Based on Appendix 1 of the Guidance on intermediates [11] completed with a helpful header.
16 Based on Appendix 3 of the Guidance on intermediates [11] completed with a helpful header.
17 Based on Appendix 2 of the Guidance on intermediates [11] completed with a helpful header.
18 See Article 3 (22) of the REACH regulation [1].
19 See Article 3 (23) of the REACH regulation [1].

6
Data Requirements and Dossier Preparation

6.1
Data Requirements

Depending on the type of registration and tonnage band to be registered the data requirements differ.

For on-site isolated intermediates no special data requirements have to be fulfilled. The same is valid for transported isolated intermediates manufactured or imported in amounts of less than 1000 t/a, whereas for a transported isolated intermediate >1000 t/a data requirements as listed in Annex VII of the REACH regulation [1] have to be fulfilled in accordance with REACH Article 18 (3) [2].

In the case of a full registration/"registration as a substance" data requirements depend on the yearly tonnage manufactured/imported by the registrant. In Article 10 of the REACH regulation [3] is listed all the information that has to be submitted with the dossier for general registration purposes.

For a full registration in the tonnage band 1 to 10 t/a data requirements in accordance with Annex VII [4] have to be fulfilled as is needed for transported isolated intermediates above 1000 t/a.

For the tonnage band 10 to 100 t/a in addition there are data requirements as listed in REACH Annex VIII [5] and furthermore for all tonnage bands above 10 t/a a Chemical Safety Report (CSR) as requested by Article 10 (b) [6] and Article 14 (1) [7] of REACH has to be submitted with the registration dossier.

For the tonnage band 100 to 1000 t/a data requirements as described in REACH Annex VII, VIII and IX [8] are relevant.

For the highest tonnage band >1000 t/a data requirements with reference to Annex VII to X [9] have to be considered.

In the case of some studies that are requested for the tonnage bands above 100 t/a, that means studies listed in Annex IX or X [10], it is possible to submit a proposal concerning this test with the registration dossier instead of doing the study before the dossier is submitted. Tests and studies requested by Annexes VII and VIII [11] in general must be done before the dossier is submitted to ECHA. There are only a few exemptions that allow deviations from the usual procedure. For a study that is needed based on Annex IX or X [10] and is appropriate to include data requirements requested by Annex VIII [5] it is possible to

REACH Compliance – The Great Challenge for Globally Acting Enterprises, First Edition.
Susanne Kamptmann.
© 2014 Wiley-VCH Verlag GmbH & Co. KGaA. Published 2014 by Wiley-VCH Verlag GmbH & Co. KGaA.

submit a registration dossier for the tonnage band above 100 t/a by including the proposal for the mentioned study and not to submit the study requested in accordance with Annex VIII [5].

In Annexes VII to X [9] there are listed the studies and tests to be done in the first column. The second column shows all reasons that are appropriate to do a waiving. Waiving means that a test or study does not have to be done because of special reasons that are given in this second column.

In very few cases it is also possible to suggest another test or study instead of the one that is foreseen. In such a case, one must have arguments in accordance with Annex XI [12] of the REACH regulation.

For full registrations every dossier must contain analytical data and descriptions of the methods as listed in Sections 2.3.5 to 2.3.7 of Annex VI [13] – for intermediate dossiers the corresponding part of the dossier is not checked with regard to completeness.

A short overview on data requirements depending on type of registration and tonnage band can be found in Figure 6.1. Depending on the type of registration and tonnage band the data requirements must be fulfilled by providing appropriate data and information in a registration dossier.

6.2
Dossier Preparation

Data are submitted to ECHA in a Dossier that is prepared by using the IT-tool IUCLID (IUCLID = International Uniform Chemical Information Database) that is supplied by ECHA free of charge. It is important always to use the latest available version to avoid problems. Actually, the version IUCLID5.4 is valid and dossiers can only be submitted to ECHA when prepared by using this version.

A substance data set as the basis for the preparation of a registration dossier consists of several chapters and sections. Depending on the type of registration and the resulting data requirements only certain chapters and sections are relevant. It is not necessary to fill in every section for a certain type of dossier, but it is necessary to fill in accurately all relevant parts of the substance data set that are relevant for the preparation of the requested type of registration dossier. For on-site isolated intermediates and transported isolated intermediates analytical data and methods are not evaluated, therefore the corresponding IUCLID Section 1.4 up to now can be left empty without any harm, whereas it will be an obligation to fill in this section for full registrations or for inquiry dossiers previously to the registration of non-phase-in substances. To be on the safe side, please always check the actual situation before submitting your dossier by reading the latest version of the relevant Guidance documents, Data Submission Manuals and fact sheets published by ECHA and also by using the technical-completeness-check tool (TCC tool) within IUCLID.

Before turning towards the single types of dossiers and presentation of the IUCLID chapters to be filled in, I will present a short overview on the fourteen

	Analytical data as listed in Annex VI [13] sections 2.3.5 to 2.3.7	Annex VII [4]	Annex VIII [5]	Annex IX [10]	Annex X [10]	Chemical Safety Report (CSR)	Remarks
Registration of an onsite-isolated intermediate (OII)	-	-	-	-	-	-	Only available data to be included
Registration of a transported isolated intermediate below 1000 t/a	-	-	-	-	-	-	Only available data to be included
Registration of a transported isolated intermediate > 1000 t/a	-	x	-	-	-	-	*)
Standard registration, 1 to 10 t/a	x	x	-	-	-	-	*)
Standard registration, 10 to 100 t/a	x	x	x	-	-	x	For full registrations above 10 t/a a CSR is needed, *)
Standard registration, 100 to 1000 t/a	x	x	x	x	-	x	For studies concerning Annex IX and X proposals can be submitted
Standard registration, > 1000 t/a	x	x	x	x	x	x	

Figure 6.1 Data requirements (checked by ECHA) depending on type registration and tonnage band.
*) When further studies/tests are available they shall be included.

IUCLID chapters existing now in Figure 6.2. Furthermore, the most important sections are considered.

If several companies cooperate for the purpose of joint registration, only the Lead Company has to enter data into Chapters 2 and 4 to 7 and submit these information in their registration dossier also on behalf of the members of the Joint submission. When any other member of the Joint submission fills in data in one or several of these chapters ECHA will assume that the company doing so is interested in doing an Opt-out for the requested parts of the dossier. As an Opt-out is only allowed in a few cases and in any case must be justified, Members of Joint submission normally do not intend to opt-out.

IUCLID Chapter	Title/Content in IUCLID Version 5.4
1	**General information**
1.1	Identification
1.2	Composition
1.3	Identifiers
1.4	Analytical information
1.5	Joint submission
1.6	Sponsors
1.7	Suppliers *)
1.8	Recipients
1.9	Product and process oriented research and development
2	**Classification & Labelling and PBT assessment**
2.1	GHS
2.2	DSD - DPD
2.3	PBT assessment
3	**Manufacture, use and exposure**
3.1	Technological process
3.2	Estimated quantities
3.3	Sites
3.4	Information on mixtures
3.5	Lifecycle description
3.6	Uses advised against
3.7	Exposure scenarios, exposure and risk assessment
3.8	Biocidal information
3.10	Application for authorisation of uses
4	**Physical and chemical properties**
5	**Environmental fate and pathways**
6	**Ecotoxicological information**
7	**Toxicological information**
8	**Analytical methods**
9	**Residues in food and feedingstuffs**
10	**Effectiveness against target organisms**
11	**Guidance on safe use**
12	**Literature search**
13	**Assessment Reports **)**
14	**Information requirements**

Figure 6.2 The IUCLID chapters and their most important sections.
*) When the registrant acts as an OR for a Non-EU manufacturer here the Appointment Letter/Only Representative Certificate has to be attached.
**) In Chapter 13 either the CSR is attached or a document giving a proof for the correctness of registration as a transported isolated intermediate or an on-site isolated intermediate may be attached (see Chapter 5 of this book).

In Figure 6.3 you will find in detail the 23 sections of IUCLID Chapter 4 Physical and chemical properties. Each of these sections corresponds to one endpoint.

In Figure 6.4 you can find details concerning the sections in IUCLID Chapters 5 to 7.

Conditions that have to be considered in entering data and information into a IUCLID Substance Data Set are in short as follows:

IUCLID Chapter	Title/Content in IUCLID Version 5.4
4	**Physical and chemical properties**
4.1	Appearance/physical state/colour
4.2	Melting point/freezing point
4.3	Boiling point
4.4	Density
4.5	Particle size distribution (Granulometry)
4.6	Vapour pressure
4.7	Partition coefficient
4.8	Water solubility
4.9	Solubility in organic solvents/fat solubility
4.10	Surface tension
4.11	Flash point
4.12	Auto-flammability
4.13	Flammability
4.14	Explosiveness
4.15	Oxidising properties
4.16	Oxidation reduction potential
4.17	Stability in organic solvents and identity of relevant degradation products
4.18	Storage stability and reactivity towards container material
4.19	Stability: thermal, sunlight, metals
4.20	pH
4.21	Dissociation constant
4.22	Viscosity
4.23	Additional physico-chemical information

Figure 6.3 Sections in the IUCLID Chapter 4 Physical and chemical properties.

- type of submission (PPORD, Inquiry, registration);
- Single submission, Lead company or Member of Joint submission;
- Role of the registrant itself M and/or I or OR;
- Type of registration (standard, TII, OII);
- tonnage band (see Figure 6.1) (CSR for 10 t/a and above, for higher tonnage bands more data have to be filled in Chapters 4 to 7).

In the next sections of this book you will find short overviews on information requirements depending on the type of dossier to be submitted to ECHA and depending on the role of the registrant. You will find fewer chapters need to be filled in for a Member of Joint submission compared to the dossier for the Lead registrant as the Lead registrant will submit parts of the dossier and maybe the CSR and Guidance on safe use on behalf of all members of the Joint submission. When a company as Member of Joint submission submits a chapter that they are not obliged to submit, because in general it is provided by the Lead Company in their registration dossier, which indicates that the Member of Joint submission wants to do an opt-out at least for this chapter. There can be good reasons for doing an opt-out, but please be aware that in depth evaluation of such dossiers by ECHA will highly likely occur and a reasonable justification for the opt-out has to be provided in the dossier.

IUCLID Chapter	Title/Content in IUCLID Version 5.4
5	**Environmental fate and pathways**
5.1	Stability
5.2	Biodegradation
5.3	Bioaccumulation
5.4	Transport and distribution
5.5	Environmental data
5.6	Additional information on environmental fate and behavior
6	**Ecotoxicological information**
6.1	Aquatic toxicity
6.2	Sediment toxicity
6.3	Terrestrial toxicity
6.4	Biological effects monitoring
6.5	Biotransformation and kinetics
6.6	Additional ecotoxicological information
7	**Toxicological information**
7.1	Toxicokinetics, metabolism and distribution
7.2	Acute toxicity
7.3	Irritation/corrosion
7.4	Sensitisation
7.5	Repeated dose toxicity
7.6	Genetic toxicity
7.7	Carcinogenicity
7.8	Toxicity to reproduction
7.9	Specific investigations
7.10	Exposure related observations in humans
7.11	Toxic effects on livestock and pets
7.12	Additional toxicological information

Figure 6.4 Sections in IUCLID Chapters 5 to 7.

A spreadsheet showing the IUCLID chapters to be filled in for each case is provided. The spread-sheet can be used as a Check-List when you intend to prepare several IUCLID dossiers in parallel or to gain some experience on how much time is needed to prepare the single chapters. It is also very useful for the last examination of your dossier before submission to ECHA.

For the sake of completeness there are also sections for the preparation of a PPORD dossier and for an Inquiry Dossier.

The PPORD dossier can be submitted when the substance is exempted from the general registration obligation for the time being. Conditions that have to be met for a PPORD notification are described in detail in Chapter 5 of this book.

An Inquiry Dossier has to be submitted previous to the registration dossier for non-phase-in substances, when you did not pre-register a phase-in substance for that you cannot do a late pre-registration any longer or when the tonnage band increases if your company needs further studies. An Inquiry Dossier can be prepared either in IUCLID or it can be done directly online in REACH-IT. Here, only the option of the preparation in IUCLID is considered.

For every Inquiry Dossier it is required to fill in the IUCLID Section 1.4 containing analytical data and methods even when the mentioned substance will be registered as an intermediate later without having further evaluation of this data requirements for the registration dossier itself.

6.2.1
PPORD

Who can submit a PPORD notification?

- M, I or OR

When to submit a PPORD notification?

- When a substance is manufactured, imported or used for the purpose of scientific research and development in a quantity of less than 1 t/a it is exempted from registration obligations. From 1 t/a and above a PPORD notification may be appropriate.
- Scientific development on a substance consisting of, for example, campaign(s) for the scaling-up or improvement of a production process in a pilot-plant or in the full-scale production [14] (irrespective of the tonnage involved].
- Investigation of the fields of application for the substance (irrespective of the tonnage involved) [14].
- Notification must be done at least two weeks before manufacture/import [15].

Data requirements [16]

- analytics in accordance with REACH Annex VI, Sections 2.3.5 to 2.3.7;
- classification, if any [16];
- there is no need to provide data for Chapters 4 to 13, but additional information can be provided.

Documents to be attached to the dossier

- spectra, chromatograms and description of analytical methods;
- in case of OR the Only Representative Certificate.

IUCLID5 chapters and sections to be filled in

- Chapter 2, Sections 3.1, 3.2, 3.4, 3.5, 3.6, 3.7, Chapters 11 and 13 will not be checked for completeness [17].
- Sections 1.8 and 1.9 will be checked for completeness [17].
- There is no need to provide IUCLID Chapters 4 to 7 as it is required in case of registrations.
- Confidentiality requests are possible if indicated.
- for Manufacturers or Importers Section 3.3 must be filled in (either manufacturing or use site).
- for an OR in Section 1.7 the Only Representative Certificate has to be attached. An overview on the relevant sections to be entered in the format of a Check-list can be found in Figure 6.5.

Check-List PPORD				
Substance name	**SUBSTANCE1**			
CAS				
EINECS				
IUCLID Chapter	To be entered by M	To be entered by I	To be entered by OR	Already entered
1.1	x	x	x	✓
1.2	x	x	x	✓
1.3	- *)	- *)	- *)	-
1.4	x	x	x	
1.5	-	-	-	
1.6	-	-	-	
1.7	-	-	x	
1.8	x	x	x	✓
1.9	x	x	x	
2.1	x**)	x**)	x**)	
2.2	-	-	-	
2.3	-	-	-	
3.1	-	-	-	
3.2	-	-	-	
3.3	x	x	-	
3.4	-	-	-	
3.5	-	-	-	
3.6	-	-	-	
3.7	-	-	-	
3.8	-	-	-	
3.10	-	-	-	
11	-	-	-	
13	-	-	-	

Figure 6.5 Check-List for Preparation of the Substance Data Set in IUCLID5 for a PPORD notification.
*) In Section 1.3 has to be provided a REACH PPORD notification number in the case of an update of a PPORD notification.
**) Will not be checked for completeness.

6.2.2
Inquiry Dossier

When to submit an Inquiry Dossier?

- for non-phase-in substances previously to the submission of the registration dossier;
- for phase-in substances that not have been pre-registered and for which a late pre-registration is no longer possible;
- before updating a dossier for already registered substances when the update has to be done because of tonnage band increase <u>and</u> there are further studies needed (not necessary if there are no further studies needed by the registrant).

Types of Inquiry (relevant type must be chosen when preparing the dossier)

- Type 1: Inquiry for non-phase-in substance.
- Type 2: Inquiry for non-phase-in substance legally on the market before June 2008.
- Type 3: Inquiry for phase-in substance that has not been pre-registered.
- Type 4: Inquiry for tonnage band increase.

Data requirements

- In any case, analytical data and methods in accordance with Annex VI, Sections 2.3.5. to 2.3.7 [13].

Documents to be attached to the dossier

- Spectra, chromatograms and description of analytical methods to be attached in Section 1.4.
- When the Inquiry Dossier is submitted by an Only Representative: Appointment Letter/Only representative Certificate has to be attached to the Dossier in Section 1.7.

IUCLID sections to be filled in

- 1.1, 1.2, 1.3 (only in case the substance already has been registered), 1.4, 1.7 (in case of an OR acting on behalf of a Non-EU manufacturer), 3.3 (for manufacturer, importer within the EU), 14 [18]. A corresponding Check-list is provided in Figure 6.6.

6.2.3
On-site Isolated Intermediate

Who can submit a registration for an OII?

- M (located within the EU).

When to submit a registration dossier for an on-site isolated intermediate?

- Only relevant for manufacturers within the EU when a substance is manufactured and afterwards isolated and stored before using this substance for transforming into another substance on the same site.
- Will never be relevant for Non-EU manufacturers, therefore neither an OR nor any importer will register an OII.

Data requirements

- Only available data has to be provided (literature research has to be done in any case).

Check-List Inquiry			
Substance name	SUBSTANCE 1		
CAS			
EINECS			
IUCLID Chapter	To be entered by M, I	To be entered by OR	Already entered
1.1	x	x	✓
1.2	x	x	
1.3	x*)	x*)	
1.4	x	x	
1.5	-	-	
1.6	-	-	
1.7	-	x **)	
1.8	-	-	
1.9	-	-	
3.1	-	-	
3.2	-	-	
3.3	x	-	
11	-	-	
14	x	x	

Figure 6.6 Check-List for Preparation of the Substance Data Set in IUCLID5 when it is intended to submit an inquiry dossier.
*) Only for an Update for an already registered substance shall be submitted this section can be filled in – in this case the registration number shall be provided.
**) If the Inquiry Dossier is submitted by an Only Representative acting on behalf of a Non-European Manufacturer, here the Appointment Letter/Only representative Certificate has to be attached (an example of such a document can be found in Figure 6.13 of this book).

- There is no need to add robust study summaries when there are no studies available – Lead company will provide as much information as available in Chapters 4 to 7.
- Lead Company will provide Chapter 2.

Documents to be attached to the dossier

- Document in which it is confirmed that the substance is manufactured and used under strictly controlled conditions may be attached in IUCLID Chapter 13 (see also Section 5.3 of this book and Appendix 3 of the Guidance on intermediates [19]). Actually this section is not checked for completeness.

IUCLID sections to be filled in

- 1.1, 1.2, 1.3, 3.1 (standard phrase), 3.3, 3.5, 11, 13 (Attachment recommended). For a Lead Company furthermore Chapter 2 is relevant.
- Remark: Sections 1.4, 3.1, 3.2, 3.4, 3.6 and 13 will not be checked for completeness. A Check-list is provided in Figure 6.7 for a company doing a single submission or for a Lead company, whereas the situation for a member of joint submission is covered in the Check-list in Figure 6.8.

Check-List OII, Lead Company or single submission		
Substance name	**SUBSTANCE 1**	
CAS		
EC		
IUCLID Chapter	To be entered by M	Already entered
1.1	x	
1.2	x	
1.3	x	
1.4	-	
1.5	-	
1.6	-	
1.7	-	
1.8	-	
1.9	-	
2.1	x *)	
2.2	-	
3.1	x **)	
3.2	-	
3.3	x	
3.4	-	
3.5	x	
4	provide as much data as available	
5		
6		
7		
11	x	
13	RMM_details, document may be attached (so far not checked for completeness)	

Figure 6.7 Registration of an OII–Check-List for the Lead Company (Manufacturer within the EU).
*) Only for a single submission/for the Lead Company in the case of Joint submission or when a company as a Member of Joint submission wants to do an Opt-out concerning this chapter.
**) Standard phrase indicating that the substance is manufactured in accordance with Article 17 (3) [20].

6.2.4
Transported Isolated Intermediate

Who can submit a registration for a transported isolated intermediate?

- M, I or OR.

When to submit a Registration Dossier for a transported isolated intermediate?

- substance is manufactured and used for the synthesis of another substance in accordance with REACH Article 17 (3) and Article 18 (4) [21];
- manufacturing and use may occur on different sites and therefore the substance may be transported.

Check-List OII, Member of Joint submission		
Substance name	**SUBSTANCE 1**	
CAS		
EC		
Role of the Registrant under REACH		
IUCLID Chapter	To be entered by M	Already entered
1.1	x	✓
1.2	x	
1.3	x	
1.4	–	
1.5	–	
1.6	–	
1.7	–	
1.8	–	
1.9	–	
2.1	– *)	
3.1	x **)	✓
3.2	–	
3.3	x	
3.4	–	
3.5	x	
11	x	
13	RMM_details, document may be attached (so far not checked for completeness)	

Figure 6.8 Registration of an OII–Check-List for a Member of Joint submission.
*) Chapters 2 and 4 to 7 have only to be filled for a Member of Joint submission if the company intends to do an opt-out for such a chapter.
**) Standard phrase indicating that the substance is manufactured in accordance with Article 17 (3) [20].

Data requirements

- Registrant must have received confirmation from all customers that they always use the substance in accordance with Article 18 (4) for the synthesis of another substance. These documents must not be attached to the registration dossier, but must be stored within the registrant's company and passed on to national authorities on request.

- For transported isolated intermediate >1000 t/a data requirements in accordance with Annex VII [4], for transported isolated intermediates <1000 t/a no special data requirements

Documents to be attached to the dossier

- When the Registration Dossier is submitted by an Only Representative: Appointment Letter/Only representative Certificate has to be attached to the Dossier.

- Document in which it is confirmed that the substance is manufactured and used under strictly controlled conditions may be attached in IUCLID Chapter 13 (see also Section 5.3 of this book and Appendix 3 of the Guidance on intermediates [19]). Actually this section is not checked for completeness.

IUCLID chapters and sections to be filled in

- 1.1, 1.2, 1.3;
- 1.7 (OR Certificate in case the registrant acts as an OR);
- 2.1 is mandatory for the Lead Company;
- 3.1 standard phrase concerning manufacturing under strictly controlled conditions is recommended;
- 3.3 (for registrants acting as M or I located within the EU);
- 3.5 (only boxes for Manufacture and Industrial Use are appropriate);
- Chapters 4 to 7 depending on tonnage band and availability of data;
- Chapter 11 must be entered by each registrant in case of a TII;
- Remark: Sections 1.4, 3.1, 3.2, 3.4, 3.6 and Chapter 13 will not be checked for completeness.

6.2.4.1 Check-List for Preparation of the Substance Data Set in IUCLID5

A Check-List for the Lead Company or in Case of Single Submission is provided in Figure 6.9, whereas a

Check-List for Manufacturers, Importers and Only Representatives being Member of a Joint Submission is given in Figure 6.10.

6.2.5
Standard Registration (Full Registration)

When to submit a Dossier for standard registration?

- for all substances with end use, for example, consumer use, for intermediates when they are not manufactured/handled under strictly controlled conditions;
- for substances used as solvent;
- for catalysts and monomers used in manufacturing of polymers in the meaning of REACH.

Data requirements

- In any case analytical data and methods in accordance with Annex VI 2.3.5 to 2.3.7 [13].
- Data requirements as listed in Annex VII to X [9] depending on the tonnage band.

Documents to be attached to the dossier

- Spectra, chromatograms, further analytical data and description of analytical methods.

Check-List Registration as TII for the Lead Registrant				
Substance name		SUBSTANCE 1		
CAS				
EC				
IUCLID Chapter	To be entered by M	To be entered by I	To be entered by OR	Already entered
1.1	x	x	x	
1.2	x	x	x	
1.3	x	x	x	
1.4	-	-	-	
1.5	-	-	-	
1.6	-	-	-	
1.7	-	-	x	
1.8	-	-	-	
1.9	-	-	-	
2.1	x	x	x	
2.2	-	-	-	
3.1	x **)	x **)	x **)	
3.2	-	-	-	
3.3	x	x	-	
3.4	-	-	-	
3.5	x	x	x	
3.6	-	-	-	
3.7	-	-	-	
3.8	-	-	-	
3.9	-	-	-	
3.10	-	-	-	
4	x	x	x	
5	x	x	x	
6	x	x	x	
7	x	x	x	
11	x	x	x	
13	RMM_details may be attached	RMM_details may be attached	RMM_details may be attached	

Figure 6.9 Check-List for Manufacturers and Importers located in the EU acting as Lead Registrant or for companies submitting a single submission.
**) Standard phrase indicating that the substance is manufactured in accordance with Article 17 (3) and Article 18 (4) [21].

- When the registration dossier is submitted by an Only Representative: Appointment Letter/Only representative Certificate has to be attached to the Dossier.
- CSR for tonnage bands above 10 t/a (at least for the Lead Registrant).

IUCLID chapters and sections to be filled in

- Sections 1.1, 1.2, 1.3, 1.4.
- Chapters 2 and 4 to 7 (to be done by the Lead company).
- Sections 3.1, 3.2, 3.3, 3.5. Sections 3.1 and 3.3 can be left empty when an OR is the registrant as the manufacturing process and site of the Non-EU

Check-List for registration as TII for M, I or OR as a Member of Joint submission				
Substance name	**SUBSTANCE 1**			
CAS				
EC				
IUCLID Chapter	To be entered by M	To be entered by I	To be entered by OR	Already entered
1.1	x	x	x	
1.2	x	x	x	
1.3	x	x	x	
1.4	–	–	–	
1.5	–	–	–	
1.6	–	–	–	
1.7	–	–	x	
1.8	–	–	–	
1.9	–	–	–	
3.1	x **)	x **)	x **)	
3.2	–	–	–	
3.3	x	x	–	
3.4	–	–	–	
3.5	x	x	x	
3.6	–	–	–	
3.7	–	–	–	
3.8	–	–	–	
3.9	–	–	–	
3.10	–	–	–	
11	x	x	x	
13	RMM_details may be attached	RMM_details may be attached	RMM_details may be attached	

Figure 6.10 Check-List for Members of Joint submission.
Remark: Chapters 2 and 4 to 7 have only to be filled in by the Lead Company.
**) A sentence or phrase indicating that the substance is manufactured and used under strictly controlled conditions in accordance with article 18 (4) of the REACH regulation. This sentence can be filled in instead of providing a detailed manufacturing process.

manufacturer are not relevant in matters of REACH, because of the OR taking over the obligations of an importer in terms of REACH. Section 3.1 can be left empty for an Importer, but Section 3.3 shall be filled in as the use takes place within EU.

- Chapter 11 (Guidance on safe use) may be done by the Lead company on behalf of all registrants that do a standard registration, but can also be done by each registrant on his own. Members of joint submission that register only a TII have to provide their own Guidance on safe use in Chapter 11.

- Chapter 13 – from 10 t/a a CSR needs to be attached at least by the Lead company, Members of Joint submission may attach their own CSR (complete or parts of it). A Check-list for Lead registrants is given in Figure 6.11, Members of Joint submission are provided with a corresponding Check-list in Figure 6.12.

Check-List for standard registration, Lead Registrant				
Substance name		***SUBSTANCE 1***		
CAS				
EC				
IUCLID Chapter	To be entered by M	To be entered by I	To be entered by OR	Already entered
1.1	x	x	x	
1.2	x	x	x	
1.3	x	x	x	
1.4	x	x	x	
1.5	-	-	-	
1.6	-	-	-	
1.7	-	-	x	
1.8	-	-	-	
1.9	-	-	-	
2.1	x	x	x	
2.2	-	-	-	
2.3	x	x	x	
3.1	x	x	x	
3.2	x	x	x	
3.3	x	x	x	
3.4	-	-	-	
3.5	x	x	x	
3.6	-	-	-	
3.7	-	-	-	
3.8	-	-	-	
3.9	-	-	-	
3.10	-	-	-	
4	x	x	x	
5	x	x	x	
6	x	x	x	
7	x	x	x	
11	x	x	x	
13	x (CSR)	x (CSR)	x (CSR)	

Figure 6.11 Check-List for the Lead Company or in case of single submission for a standard registration (full registration).

6.3
Some Useful Tips for Entering Data and Information in Certain Chapters in IUCLID5.4

6.3.1
IUCLID Section 1.2

Degree of purity, typical concentration and concentration range shall be provided in the unit "% w/w". Information on the typical concentration and concentration range is expected for the main constituent(s) and all relevant impurities. In sum, the typical concentration of main constituent(s) and impurities must be 100.0%.

Check-List for full registration, M/I, Member of Joint submission				
Substance name		SUBSTANCE 1		
CAS				
EC				
IUCLID Chapter	To be entered by M	To be entered by I	To be entered by OR	Already entered
1.1	x	x	x	
1.2	x	x	x	
1.3	x	x	x	
1.4	x	x	x	
1.5	–	–	–	
1.6	–	–	–	
1.7	–	–	x	
1.8	–	–	–	
1.9	–	–	–	
2.3	x	x	x	
3.1	x	x	x	
3.2	x	x	x	
3.3	x	x	–	
3.4	–	–	–	
3.5	x	x	x	
3.6	–	–	–	
3.7	–	–	–	
3.8	–	–	–	
3.9	–	–	–	
3.10	–	–	–	
11	x*)	x*)	x*)	
13	x*)	x*)	x*)	

Figure 6.12 Check-List for Manufacturers, Importers and Only Representatives within the EU being a Member of Joint submission.
Chapters 2 and 4 to 7 have to be filled by the Lead Company on behalf of the Members of Joint submission. Fill Members such a chapter than it means that they opt-out concerning this Chapter.
*) Chapter 11 and the CSR in 13 can be done by the Lead Company on behalf of the Members or all Members submit the information on their own.
Section 2.1 and Chapters 4 to 7 have to be filled by the Lead Company on behalf of the Members of Joint submission. If a Member fills in such a chapter that means that they opt-out concerning this chapter.

Furthermore, it is expected that figures for the typical concentration correlate with the values given in chromatograms attached in Section 1.4.

6.3.2
IUCLID Section 1.3

In the registration dossier of a previously pre-registered substance it is expected that you provide ECHA with the pre-registration number that your company received for this substance. The format of the pre-registration number is

05-xxxxxxxxxx-xxxx-xx (pre-registration done until 30.11.2008)

17-xxxxxxxxxx-xxxx-xx (late pre-registration done after 30.11.2008, because of first-time manufacturing or import).

For a non-phase-in substance instead the Inquiry number has to be entered in the format 06-xxxxxxxxxx-xxxx-xx.

If it is intended to benefit from an extension of a PPORD notification, in this chapter the PPORD notification number shall be stated in the format 04-xxxxxxxxxx-xxxx-xx.

When an update of an already registered substance shall be submitted at least the registration number in the format 01-xxxxxxxxxx-xxxx-xx has to be provided.

Section 1.3 can only be left empty without rejection either in case of Inquiry Dossiers that are submitted previously to a registration dossier for a non-phase-in substance or in case of a PPORD notification submitted for a certain substance for the first time.

6.3.3
IUCLID Section 1.4

In dossiers for the purpose of registration of transported isolated intermediates or on-site isolated intermediates this chapter may be left empty, because it is not checked for completeness yet. In all other cases – Inquiry Dossiers, Dossiers for standard registration or in PPORD notifications – it is highly recommended to fill in this chapter with due diligence and attach all relevant documents.

ECHA expects in accordance with Annex VI, 2.3.5 to 2.3.7 [13] the following spectra and chromatograms:

- UV/Vis;
- IR;
- NMR and/or MS;
- HPLC and/or GC.

If one or several of these methods are not appropriate to determine the identity and purity of a certain substance, these techniques may be listed and a justification (Waiving argument) for not having provided data concerning the requested method has to be given. For inorganic substances other methods may be provided instead of the usually requested ones. It is important to choose analytical methods to determine the identity and also the purity of the substance that is the subject of the dossier.

ECHA furthermore demands that a description of the relevant analytical methods is attached in Section 1.4. Alternatively, there could be stated a method that is available in the literature or certain guidelines. In any case, it is important that the method can be reproduced and it is expected that the method allows determination of the content of certain constituents in the unit "% w/w". Therefore, it may be necessary to provide ECHA also with calibration methods and all relevant calculations.

6.3.4
IUCLID Section 1.7

If an Only Representative as indicated in IUCLID Section 1.1 takes care of the registration obligations on behalf of a Non-EU Manufacturer in Section 1.7 an Appointment Letter or Only Representative Certificate has to be attached.

The document below (see Figure 6.13) can be used as a Template for such an Appointment letter.

Chemistry & Co. Ltd.
Oil Street 235
77865 Prisbane
Australia

Only Representative Certificate

In compliance with Regulation (EC) No 1907/2006, of the European Parliament and of the Council of 18 December 2006 concerning the Registration, Evaluation, Authorization and Restriction of Chemicals (REACh), the Company
Chemistry & Co. Ltd. (non-EU entity) having its principal business place at **Prisbane, Australia** has appointed *[Name of Only Representative]* (OR), having its principal place of business at *[Place, Country within the EU]*, to act as its Only Representative in accordance with Article 8 of the REACH regulation, to take care of the REACH compliance process on its behalf for the following substance:

Name	Abbreviation	CAS	EC

The appointed Only Representative [name of Only Representative] is identified under the Universal Unique Identifier (UUID) code:
[UUID]

[Place, date]
[signature]
[Title, name and function]

Figure 6.13 Template for an Appointment Letter/Only Representative Certificate.

6.3.5
IUCLID Section 2.3

In the IUCLID5.4 version there is a new chapter concerning PBT Assessment.

For a standard registration all necessary information to be filled in in this chapter can be found in Chapter 8 of the CSR and just has to be copied to IUCLID Section 2.3.

But there may also occur situations that do not require a PBT assessment, for example, in the case of inorganic substances. Then, in "PBT assessment: the overall result" the option "PBT assessment does not apply" can be chosen and a justification has to be given. For inorganic substances a justification could be: "A PBT and vPvB assessment shall be done as foreseen in REACH Article 14 (3) (d) in conjunction with REACH Annex I Section 4 according to the criteria laid down in REACH Annex XIII. According to Annex XIII [22] of the REACH regulation a PBT and vPvB assessment shall be carried out for organic substances, including organo-metals. As [. . .] is an inorganic substance, a PBT and vPvB assessment is not required."

6.3.6
IUCLID Section 3.1

When the registration dossier shall be submitted either by an Importer or an Only Representative it is not necessary to provide details on the manufacturing process, whereas it is required to provide at least a short description in the case of a Manufacturer. However, it is recommended to give a short statement whenever the dossier is intended for the purpose of a registration of an intermediate. For a transported isolated intermediate in IUCLID Section 3.1 in the block "Methods of manufacture of substance" you may state:

> This substance is handled under strictly controlled conditions in accordance with REACH regulation Article 17 (3) for on-site isolated intermediates, and when the substance is transported to other sites for further processing, the substance should be handled at these sites under the strictly controlled conditions as specified in REACH regulation Article 18 (4). Site documentation to support safe handling arrangements including the selection of engineering, administrative and personal protective equipment controls in accordance with risk-based management systems is available at each Manufacturing site.
>
> Written confirmation of application of strictly controlled conditions has been received from every affected Distributor and every Downstream Manufacturer/User of the registrant's intermediate.

This standard phrase can also be used as an introduction in the blocks "Handling and storage" and "Exposure controls/personal protection" in the Guidance on safe use to be provided in IUCLID Chapter 11.

For the registration of an on-site isolated intermediate this standard phrase can easily be adapted in an appropriate way.

6.3.7
IUCLID Chapter 11

It is recommended to use also a standard phrase as given in the previous section of this book for IUCLID Section 3.1 for transported isolated intermediates. This standard phrase should be written in the beginning of the blocks "Handling and storage" and "Exposure controls/personal protection" to prove that the substance to be registered fulfills the criteria for a transported isolated intermediate.

The same standard phrase may be used as a justification in the dossier header when creating the dossier for a transported isolated intermediate.

For a Joint submission it is not demanded that all Members submit the same text concerning the Guidance on safe use, but for the sake of consistency it may be useful to find an agreement among all Members of the Joint submission.

6.3.8
IUCLID Chapter 13

For a registration as an intermediate a document in which it is confirmed that the substance is manufactured and used under strictly controlled conditions may be attached (see also Section 5.3 of this book and Appendix 3 of the Guidance on intermediates [19]). Actually this section is not checked for completeness.

For a standard registration above 10 t/a a Chemical Safety Report is required, this CSR may be submitted by the Lead Company on behalf of all Members of the Joint submission, then Members of a Joint submission do not need to attach a CSR document in Chapter 13 but have to clearly state this situation in the dossier header.

A CSR can also be submitted in parts jointly and another part has to be submitted by each Member of the Joint submission on its own, then it shall be declared as "part CSR". If the complete CSR is submitted by each Member of the Joint submission on its own "own CSR" has to be declared.

Please check the situation for your substance and your special requirements in each case with due diligence.

6.4
Data Requirements, Type of Registration and Costs/Fees

Fees that are invoiced by ECHA are based on the type of registration. When a company submits a dossier as a Member of Joint submission the company will save about 25% of the fees compared to a single submission.

Fees are also reduced for small and medium sized companies [23].

Concerning the following overview (Figure 6.14 and Figure 6.15) it is assumed that the company is large (>250 persons working within the mentioned company) and the assumption is also made that the registration is submitted directly in the stated tonnage band. When a company does at first a registration in a smaller tonnage band and afterwards submits an update because of tonnage band increase, only the difference to the already paid fees has to be collected by ECHA.

It is foreseen that ECHA can increase the fees with reference to inflation, therefore it maybe that there are further amendments in future.

	Single submission	Joint submission
OII	1714 €	1285 €
TII	1714 €	1285 €
Standard registration, 1 to 10 t/a	1714 €	1285 €
Standard registration, 10 to 100 t/a	4605 €	3454 €
Standard registration, 100 to 1000 t/a	12 317 €	9237 €
Standard registration, > 1000 t/a	33 201 €	24 901 €

Figure 6.14 Fees for submission of Registration dossiers – assumption: large company (above 250 members of the staff) [24].

Standard fee	536 €
Extension (further 5 or 10 years)	1071 €

Figure 6.15 Fees for PPORD notification [25].

If you would like to have more details, especially concerning reduced fees that medium, small and microenterprises can benefit from, please do not hesitate to look through the fees regulation [23].

It is very clear and transparent at which level fees have to be paid with ECHA, but very often costs that a company has to take into account for the preparation of the dossier are much higher than the fees for submitting the dossier.

There will occur costs within the company intending to do a registration and there will also be cash-out.

Costs for studies and tests have to be budgeted and also costs for a consultant or some other experts (e.g., lawyers, trustees, financial management for consortia, etc.) have to be considered.

The IT-tool IUCLID5 is available from ECHA free of charge, but maybe you will need support from an IT expert for installation and maintenance work.

If you have a cooperation with other companies with the aim to prepare a Joint submission several costs (e.g., for tests, studies, consultant) can be shared, but other costs will arise (trustee, consortium management and/or SIEF management, financial management, lawyers, IT-tools as a platform for exchange of information within a consortium or SIEF, etc.).

The experience shows that SIEF management and consortium management can cause costs that are even higher than the costs for all the studies and tests to be done for a full registration.

Therefore, every company that will have to register a certain substance to be sold to customers has to answer the question whether future business will allow to take the costs that arise concerning the registration of that substance.

6.5
Examples and Exercises

1. Your company (located within the EU) is willing to register a certain acid chloride that is manufactured in amounts of 200 t/a up to 300 t/a at your site. This acid chloride shall be sold to several customers.

   a) What type of registration may be appropriate and why?

   b) What is your preferred option and why?

   c) Which data requirements have to be fulfilled?

   d) Which IUCLID5 chapters and sections have to be entered if your company is the only registrant or Lead Company?

   e) Which IUCLID5 chapters and sections are relevant if your company were to become a Member of Joint submission?

   f) Which fees will have to be paid with ECHA?

2. Assume the acid chloride from question 1 is not sold to a customer, but is used within the company that manufactured the substance at the same site, which data requirements will then be relevant?

3. What difficulties can arise when a company intends to register a non-phase-in substance?

4. Is there a limitation concerning tonnages to be handled under a PPORD notification?

5. A large company (>250 members of staff) intends to register a substance A either as a transported isolated intermediate or with a standard registration. In any case they will do a single submission.

   a) Is there any difference concerning the fees to be paid to ECHA when the registration is done in the tonnage band 1 to 10 t/a?

   b) How much is the difference if they intend to register the substance in the tonnage band 100 to 1000 t/a?

6. Which conditions have to be considered to make a decision which chapters and sections in IUCLID5 have to be filled in?

References

[1] Regulation (EC) No 1907/2006 of the European Parliament and of the Council of 18 December 2006 concerning the Registration, Evaluation, Authorization and Restriction of Chemicals (REACh); Annex VII is entitled "standard information requirements for substances manufactured or imported in quantities of 1 tonne or more".

2 See Article 18 (3) of the REACH regulation [1].
3 See Article 10 of the REACH regulation [1].
4 See Annex VII of the REACH regulation [1], Annex VII is entitled "standard information requirements for substances manufactured or imported in quantities of 1 tonne or more".
5 See Annex VIII of the REACH regulation [1].
6 See Article 10 (b) of the REACH regulation [1].
7 See Article 14 (1) of the REACH regulation [1].
8 See Annexes VII, VIII and IX of the REACH regulation [1].
9 See Annexes VII to X of the REACH regulation [1].
10 See Annexes IX and X of the REACH regulation [1].
11 See Annexes VII and VIII of the REACH regulation [1].
12 See Annex XI of the REACH regulation [1].
13 See Annex VI of the REACH regulation [1].
14 Guidance on Scientific Research and Development (SR&D) and Product and Process Oriented Research and Development (PPORD), February 2008 published by ECHA.
15 See Article 9 (5) of the REACH regulation [1].
16 See Article 9 (2) of the REACH regulation [1].
17 Data Submission Manual, Part 05 – How to complete a technical dossier for registrations and PPORD notifications, Version 3.0, 07/2012, published by ECHA.
18 Data Submission Manual, Part 02 – How to prepare and submit an inquiry dossier using IUCLID5, Version 2.0, July 2012, published by ECHA.
19 Guidance on intermediates, Version: 2, December 2010, published by the European Chemicals Agency; Guidance documents can be obtained via the website of the European Chemicals Agency (http://echa.europa.eu/reach_en.asp).
20 See Article 17 (3) of the REACH regulation [1].
21 See Articles 17 (3) and 18 (4) of the REACH regulation [1].
22 Commission Regulation (EU) No 253/2011 of 15 March 2011 amending Regulation (EC) No 1907/2006 of the European Parliament and of the Council on the Registration, Evaluation, Authorisation and Restriction of Chemicals (REACH) as regards Annex XIII.
23 COMMISSION IMPLEMENTING REGULATION (EU) No 254/2013 of 20 March 2013 amending Regulation (EC) No 340/2008 on the fees and charges payable to the European Chemicals Agency pursuant to Regulation (EC) No 1907/2006 of the European Parliament and of the Council on the Registration, Evaluation, Authorisation and Restriction of Chemicals (REACH).
24 Figures are based on Annex I Table I and Annex II Table I in the regulation on fees and charges [23].
25 See Annex V, Table I and Table II of the regulation on fees and charges [23].

7
Claiming a Registration Number for Already Notified Substances

7.1
Formerly Notified Substances are Regarded as Registered under REACH

Before REACH entered into force it was an obligation for European manufacturers to notify new substances that were marketed in accordance with Directive 67/548/EEC [1]. These substances are known as **NONS** (**no**tified **n**ew **s**ubstances) and they are not listed in EINECS. Under REACH the former "new substances" are called "non-phase-in substances". A notification had to be submitted to national authorities when a manufacturer produced or an importer imported such a substance in amounts of 10 kg/year and above. Non-European manufacturers were allowed to notify such a substance in a Member State of the EU via a Sole Representative. The function of the former Sole Representative is not identical to the Only Representative under REACH but it is at least similar.

In any case former notified substances can be regarded as registered under REACH for the party that did such a notification under Directive 67/548/EEC.

A company that did not notify a substance at the national authorities under this Directive 67/548/EEC [1] has the usual obligations under REACH that occur in the context of non-phase-in substances.

In accordance with Article 24 (1) [2] of the REACH regulation [3] a notification in accordance with Directive 67/548/EEC [1] is not only regarded as a registration under REACH, but ECHA should also assign a registration number by 1 December 2008.

The former notifier had the right – and still has the right – to claim the registration number from ECHA. In any case, the substance then will be regarded as registered in the smallest tonnage band under REACH, even if there are data gaps. After the claim of the registration number the substance is in accordance with the information that can be found in REACH-IT registered. The substance may be registered as an on-site isolated intermediate, as a transported isolated intermediate and also with a full registration, each of them for the tonnage band 1 to 10 t/a regardless of whether there are all data requirements fulfilled under REACH or not. In case there had been a notification for a higher tonnage band it is possible that the mentioned substance already will be regarded as registered in a higher tonnage band depending on the already submitted data.

REACH Compliance – The Great Challenge for Globally Acting Enterprises, First Edition. Susanne Kamptmann.
© 2014 Wiley-VCH Verlag GmbH & Co. KGaA. Published 2014 by Wiley-VCH Verlag GmbH & Co. KGaA.

If the quantity of a notified substance manufactured or imported per manufacturer or importer reaches the next tonnage threshold as described in Article 24 (2) of the REACH regulation [4], the registrant will have the obligation to submit an update. The update shall be submitted even in the case of a tonnage band increase when there is no further data requirements, for example, if the registrant wishes to register the said substance only as a transported isolated intermediate in a tonnage band below 1000 t/a.

If a registrant has to do an Update it is important to fulfill the data requirements as foreseen under REACH. If there were data gaps concerning the already registered tonnage band these gaps also have to be considered and data have to be completed with due diligence.

For the sake of completeness we will consider the situation for a manufacturer that manufactured a new substance before REACH [3] entered into force, but has not marketed this substance. For this case it was not necessary to notify the new substance in accordance with Directive 67/548/EEC [1]. Therefore, such a substance cannot be regarded as registered under REACH and the manufacturer cannot claim a registration number for this substance. However, in accordance with REACH Article 3 (20) (b) [5] this substance may have phase-in status for a certain manufacturer and he may have had the chance to pre-register this particular substance until 30 November 2008. Afterwards, the same rules concerning the registration deadline can be applied for this manufacturer as for other phase-in substances. As the substance was not marketed in former days it may be possible that the substance can be regarded as an on-site isolated intermediate.

7.2
How to Claim the Registration Number Under REACH for a Formerly Notified Substance

The registration number for a previously notified substance can be claimed by a party that manufactured or imported such a substance in former days and notified it with the national authorities. When the previous notifier had been a Sole Representative, this party in principle can act as an Only Representative and also claim the registration number, but as the Sole representative Agreements were invalid after May 31st, 2008 [6] the newly appointed Only Representative needs a new letter confirming that he is allowed to act on behalf of the Non-European manufacturer.

The claim of the registration number has to be done in REACH-IT as described in the User Manual [7]. In case an Only Representative (former Sole Representative) acts for more than one Non-EU manufacturer he has to sign up in REACH-IT for each company he represents separately and he will receive one registration number and one submission number for each of the Non-EU manufacturers he represents.

After having successfully requested a registration number, the registrant of the previously notified substance can request the original SNIF file and the IUCLID 5 migrated file from the Member State Competent Authority (MSCA) [7].

7.3
When to Update a Registration Dossier of a Formerly Notified Substance and How to Do It

In general, spontaneous Updates of the registration dossiers of previously notified substances have to be done in the same cases as for all substances that were registered under REACH [8] and also on request of ECHA. This topic will be discussed in depth in Chapter 11 of this book.

Very often the first update of the dossiers of previously notified substances was done to fulfill the obligation of CLP notification that had to be done by registrants until January 3rd, 2011 in case they bring the mentioned substance to the market.

The previously used SNIF files that national authorities placed at the previous notifiers' disposal must be converted to the actually necessary IUCLID format.

As there occur several bucks by converting the SNIF File into an IUCLID format, it was necessary that registrants checked/check the converted file with due diligence and did/do the necessary amendments to correct wrong entries. Very often the substance was for example, marked as a polymer although it is a monoconstituent substance. If a former notifier did not ask the national authorities for the relevant SNIF files in the past, he will be provided with a IUCLID5 File in the meantime. In any case it is important to have the file in an appropriate version of IUCLID as an .i5z file before it is imported to the actual version of IUCLID. When the substance data set already had been imported into an older version of IUCLID5 it will be automatically converted into the newest version of IUCLID when you upgrade your IUCLID Account.

Special rules are to be applied concerning Confidentiality flags that had been used for the previous notification. Registrants can benefit in this case from the special rule that they can keep this confidentiality flag in the dossier and do not have to pay extra fees at ECHA in the case of an update, provided that in the field "justification" adjacent to the confidentiality flag is filled in the text "Claim previously made under Directive 67/548/EEC" [6, 9]. Before submitting the dossier a quick Check with the IUCLID5 tool for estimation of the fees to be paid should be applied.

7.4
Examples and Exercises

1. What type of substance are NONS?

2. Two companies located in Germany manufactured a substance S – that was not listed in EINECS – before REACH entered into force. Company A sold this substance S to several customers whereas Company B used Substance S only at their own site.

   a) What obligation did companies A and B have concerning Directive 67/548/EEC?

   b) What actions do companies A and B have to take within the calendar years 2008/2009? Give a justification.

3. What was needed from the national authorities as a prerequisite of doing an update of the registration of a former notified substance?

4. Your company claimed a registration number for a previously notified substance that now can be regarded as registered under REACH. What type of update could be necessary at a later stage? (see also Chapter 11 of this book).

   [ ] spontaneous update

   [ ] update on request

5. What reasons may have been/may be the justification for doing a spontaneous update? (see also Chapter 11 of this book).

   [ ] CLP notification

   [ ] increase of tonnage band

   [ ] new customers

   [ ] moving of the national authority

   [ ] new knowledge of the risks of the substance to human health and/or the environment

   [ ] reasons as stated in REACH Article 22 (1) [8]

   [ ] change in terms of delivery

References

1 Council Directive of 27 June 1967 on the approximation of laws, regulations and administrative provisions relating to the classification, packaging and labeling of dangerous substances (67/548/EEC).
2 See Article 24 (1) of the REACH regulation [3].
3 Regulation (EC) No 1907/2006 of the European Parliament and of the Council of 18 December 2006, concerning the Registration, Evaluation, Authorisation and Restriction of Chemicals (REACH), establishing a European Chemicals Agency, amending Directive 1999/45/EC and repealing Council Regulation (EEC) No 793/93 and Commission Regulation (EC) No 1488/94 as well as Council Directive 76/769/EEC and Commission Directives 91/155/EEC, 93/67/EEC, 93/105/EC and 2000/21/EC.
4 See Article 24 (2) of the REACH regulation [3].
5 See Article 3 (20) (b) of the REACH regulation [3].
6 European Chemicals Agency, Questions and Answers, For the registrants of previously notified substances (Release 6), 22-09-2010.
7 European Chemicals Agency, REACH-IT Industry User Manual, Part 10 – Claim of a registration number for a notified substance.
8 See Article 22 (1) of the REACH regulation [3].
9 See Annex 3 of the Data Submission Manual, Part 05 – How to complete a technical dossier for registrations and PPORD notifications, Version 2.8, April 2011, published by the European Chemicals Agency.

8
Process for Registration of Non-Phase-In Substances

For all substances that are not considered as phase-in substances under REACH [1] the rules for non-phase-in substances have to be applied. For a quick Check, whether a substance is a phase-in substance or a non-phase-in substance the short summary in Figure 8.1 can be used.

Non-phase-in substances cannot benefit from the transitional regime provided for phase-in substances. Non-phase-in substances therefore have different timelines for registration [7]. Since 1 June 2008 (12 months after entry into force of REACH [1]) non-phase-in substances that fall under the scope of REACH (manufactured/imported in amounts of 1 t/a and above, if not exempted from REACH registration obligations) must be registered before the start of manufacture or import. The same is valid for phase-in substances that have not been pre-registered and for that also no late pre-registration can be done any longer. It is important to stress that registration of non-phase-in substances will first require the submission of an inquiry dossier [7].

Considered as Phase-in substances	**Non-Phase-in substances**

• REACH Article 3 (20) [2]
  a) listed in EINECS [3]
  b) phase-in status as manufactured within the Community but not marketed at least once in the 15 years before the entry into force of REACH [4]
  c) substance was placed on the market in the community and notified in accordance with the first indent of Article 8(1) of Directive 67/548/EEC but is not polymer [5]
• NLP [6]
	All substances that do not meet the criteria of phase-in substances in accordance with REACH Article 3 (20) and that are not NLPs

Figure 8.1 Short summary as Check-list for the decision whether a substance is a phase-in substance or a non-phase-in substance.

Inquiry Dossier to be submitted in the following cases:	Remarks
For non-phase-in substances	Same is valid for phase-in substances that have not been pre-registered and that cannot benefit from a late pre-registration any longer
For substances for which previously a PPORD notification was done, but extension of the PPORD notification is no longer possible	
Previous to the update of a registration dossier because of tonnage band increase	Only when the registrant needs further data/studies

Figure 8.2 Cases in which submission of an Inquiry Dossier previous to the registration is needed.

8.1
Inquiry Dossier

An Inquiry Dossier has to be submitted to ECHA in several cases. For the sake of completeness all cases are considered in Figure 8.2.

The submission of an inquiry dossier is demanded by ECHA in accordance with REACH Article 26 [8] to determine whether a registration or another inquiry has already been submitted for the same substance so that data sharing mechanisms can apply [7]. Studies involving vertebrate animals shall not be repeated [9]. Therefore, ECHA will inform the potential registrant, if the same substance either has previously been registered less than 12 years earlier [9] or if several potential registrants have made an inquiry in respect of the same substance [10].

ECHA will inform the potential registrant accordingly, if the same substance has previously not been registered [11].

When submitting an inquiry, potential registrants are required to submit an inquiry dossier with information as listed in Article 26 (1) [12].

Identity of the inquirer, identity of the substance of interest and also a list of which information requirements would require new studies shall be included in the dossier [12].

The identity of the inquirer will include contact details and the location of the inquirer's production site [7].

As ECHA has to evaluate based on the information given in the inquiry dossier whether the same substance has already been registered or whether there are further potential registrants the sections of the highest concern in the Inquiry Dossier will be IUCLID Section 1.2 containing the composition of the substance and Section 1.4 comprising analytical data and methods.

Based on this information ECHA will do a Sameness Check.

The information required is identical to that required in the technical dossier for registration. It is important to note that for substances used as intermediates, the information to be provided in the inquiry dossier for the identification of the substance will have to comply with the same requirements as for non-intermediates [7]. The registrant cannot benefit from reduced requirements even if he intends to register the substance later on as a transported isolated intermediate or as an on-site isolated intermediate [7].

Detailed information how to fulfill the data requirements and some tips for filling in the IUCLID Substance Data set can be found in Chapter 6 of this book.

There is no time frame defined in REACH [1] for the manual check of an inquiry dossier by ECHA staff, therefore it may last several weeks up to months from the date of submission and after having passed the Enforce rules Check until you will receive a comment from ECHA. The dossier may be rejected even when the technical completeness Check had been passed, if ECHA staff are not able to identify the substance without any doubt, because of missing information concerning analytical data and methods. In such a case you may amend the Substance Data Set in IUCLID, add missing information, create a new dossier and submit it again to ECHA. If an inquiry dossier was rejected the next submission may be done without referring to the previous trial unless ECHA defined something else in their communication.

If your inquiry dossier was considered to be prepared with due diligence, ECHA will reward this with an Inquiry number of the format 06-xxxxxxxxxx-xx-xxxx and they will also provide you with an EC list number for your substance. Both numbers are needed for the preparation of the registration dossier. As an answer on the successful submission of an inquiry dossier you will be furthermore informed whether there is already a registrant for this substance or maybe some other potential registrants. Depending on this information the further procedure for the preparation of the registration dossier will differ.

8.2
Preparation of the Registration Dossier

Based on the outcome of the inquiry there may be three different options for the further procedure, as shown in Figure 8.3. The substance may already be registered by another company. Then this company may act as the Lead Company. If there is no registrant yet, but at least one other company is willing to register the same substance there will be the obligation to cooperate, whereas if there is no other (potential) registrant the inquirer may proceed without having the obligation to cooperate with competitors or competitors to be.

Further procedure, time frame, costs for dossier preparation and fees with ECHA may differ depending on the situation that occurs after the outcome of the inquiry.

For Option 1 and Option 2 the further steps in the registration process will be similar to the situation that is described for the registration of previously pre-registered phase-in substances that are either already registered by a Lead Company (as in Option 1) or that shall be registered jointly by several companies that

8 Process for Registration of Non-Phase-In Substances

	OPTION 1	OPTION 2	OPTION 3
Outcome of the inquiry	Substance was already registered by another company	Substance was not registered yet, but there is at least one other company that has submitted an inquiry	Substance was not registered yet and there are no other potential registrants yet
Further tasks for the company that submitted the inquiry	Ask Lead company for LoA to their dossier	Find an agreement with other potential registrant(s) who will act as the Lead Company	Prepare registration dossier
Type of registration dossier template to be used by the company that submitted the inquiry	Registration as a Member of Joint submission	Joint submission – either registration as the Lead company or as a Member of Joint submission	Single submission
Costs for dossier preparation	Costs for LoA	Depending on contract(s) with other potential registrant(s)	Costs for preparation of the complete dossier, maybe costs for tests and studies if required
Time frame for preparation of the registration dossier until submission to ECHA	Can be realized within a short time after having received submission name and token by the Lead company	May last longer as the inquirer has to find an agreement concerning several tasks with the other potential registrant(s)	Depending on type of registration (standard registration, TII or OII) Can be realized within a short time for OII and TII < 1000 t/a
Fees with ECHA	Joint submission (25% less than for single submission)	Joint submission (25% less than for single submission)	Single submission
Remarks	Can be the fastest way to do the registration in case the Lead company supports the inquirer immediately. As the inquirer has to negotiate with a direct competitor it may be difficult to find an agreement concerning contracts for LoA and costs to be paid by the inquirer.		Time frame and costs can be influenced in the best possible way by the inquirer itself.

Figure 8.3   Outcome of an inquiry.

pre-registered the same substance and therefore were grouped by ECHA into the same pre-SIEF. In both cases the time needed for negotiations with registrant/potential registrants will be the bottleneck in the registration process.

For the Option 3 the inquirer has to prepare the registration dossier on his own. Depending on whether there are still needed some studies/tests that have to be done previously to the submission of the dossier it may last longer to prepare the dossier.

8.2.1
Registration as Member of Joint Submission

If the substance of concern is already registered by another company, this company may act as the Lead Registrant. The inquirer is provided with name, address and all other necessary details that are required to get into contact with the Lead Company. The Lead Company itself will also be informed by ECHA and they are provided with the name and contact details of the inquirer.

This enables the inquirer to get into contact with the Lead Company very rapidly. Normally, the Lead Company will answer soon on the request of the inquirer. ECHA assumes that both parties will be in the position to find an agreement within one month [13]. As in each case the inquirer has to cooperate with a competitor and *vice versa* the Lead Company has to face the fact that there will be a competitor in future, further negotiations in reality may be time consuming. Costs for a Letter of access that will allow the inquirer to refer to the registration dossier of the Lead Company shall be shared in a fair, transparent and non discriminatory way. Reality shows that costs for a certain type of dossier can vary over a broad range from almost nothing up to a fortune depending on the Lead Company and involved service providers.

As soon as the registrant to be is willing to accept the conditions of the Lead Company, registration can be done within a short time. The registrant to be needs submission name and token from the Lead Company to become a Member of Joint submission in REACH-IT. Afterwards, the registrant to be has to prepare a registration dossier comprising of all necessary data and information that has to be provided for a certain type of registration. Details concerning data requirements and dossier preparation can be found in Chapter 6 of this book.

Fees with ECHA will depend on the type of registration and size of the registrant's company, but will be in any case 25% less than for the corresponding single submission.

8.2.2
Registration within a Joint Submission in Cooperation with Other Potential Registrant(s)

If the outcome of an inquiry is that the substance was not yet registered, but that there is at least one other company intending to register the same substance, ECHA will provide both inquirers with the contact details of the other party. Furthermore, they will be provided with the information on studies that ECHA knows of and that are older than 12 years. Potential registrants may refer to such studies without paying for them. However, the registrants to be are obliged to agree on the further procedure. One company shall become the Lead Company, the other party will be a Member of Joint submission. Concerning a contract that can be signed by all registrants to be there are the same options as in the case of SIEFs that have in common that all members pre-registered the same phase-in substance and afterwards can prepare the registration dossier jointly. There can be a contract between the parties sharing work-load and costs on an equal basis, as is determined in a Consortium Agreement or SIEF Agreement and that is open to other parties that may register the same substance at a later stage, but there can be made also other contracts that are based on a Letter of Access model. A Letter of Access is normally signed by the Lead Company and the party that asked for the LoA.

It will take a certain time to find an agreement on who will be the Lead Company and in regard to the further procedure. Therefore, especially in cases where further studies/tests are needed the overall time from having received the outcome of the inquiry until successful submission of the registration dossier can be expected to last much longer than in the case of Options 1 or 3, as described in Figure 8.3.

Depending on the contractual arrangements the influence of each single registrant participating in the Joint submission on the diverse actions to be done and also the overall costs may be higher than when there is already a registrant that tries to earn some money by selling LoAs. The cost sharing can be expected to be fair and more transparent as the registrants to be will start their cooperation at an early stage and share costs really on an equal basis.

As each registrant will be a Member of Joint submission concerning the costs for the fees with ECHA each company can benefit from a 25% deduction compared to a single submission.

8.2.3
Single Submission

If the outcome of an Inquiry had been that there is no registrant yet and no other party intending to register the same substance the inquirer has to face the fact that he will have to do the preparation of the registration dossier on its own.

There will be no time-consuming discussions with other registrants to be concerning contractual arrangements, the procedure for preparation of the dossier or votes concerning the next steps to be done.

However, the costs for the dossier preparation cannot be shared with others for the time being and fees with ECHA have to be paid in accordance with a single submission (individual submission) without any deduction.

The time frame from outcome of the inquiry until submission of the registration dossier will depend on the type of registration dossier and the necessity of doing further tests and studies. For transported isolated intermediates below 1000 t/a the registration dossier can be prepared within a very short time, because there are no special data requirements concerning studies.

REACH	Industry
Protection of human health and the environment	✓
Avoid unnecessary testing	✓
Support the thought of "one substance one registration" by bringing together registrants and potential registrants in the inquiry process.	Often it is clever to avoid competitors being informed at an early stage about intentions of a potential registrant to come to market. Not possible if an Inquiry has to be submitted.
Does not support acting confidential until a substance is brought to market	Confidentiality sometimes is even demanded by the competition law
Does not take care of the costs registrants have to defray	Economic aspects are very important. REACH may not strengthen the competitiveness of Community industry although this is intended by REACH.

Figure 8.4 Demands of REACH versus needs of industry acting in the market.

8.3
Difficulties and Problems that can Arise in the Context of the Registration of Non-Phase-In Substances

REACH should ensure a high level of protection of human health and the environment [1] and also avoid unnecessary testing especially when there are vertebrate animals involved. The latter means that potential registrants shall cooperate in the preparation of a Joint submission as it is intended to have only one registration per substance. Although industry is also interested in the protection of human health, especially but not limited to the health of workers, the obligation to cooperate with competitors is a great challenge especially in the case of non-phase-in substances being the subject of registration obligations. Figure 8.4 shows some aspects where demands of REACH can have a negative influence on the strong need of industry to survive in a merciless market and earn money to ensure that workers within the EU can have their jobs although within the next year.

Starting a manufacturing campaign for amounts of 1 t/a and above of a non-phase-in substance within the EU is not allowed until this substance has been registered. As upscaling from lab scale to large volume production scale very often is done without manufacturing in the meantime several kg amounts in a pilot campaign the potential registrant will have difficulties in providing ECHA with information concerning composition and analytical data based on the quality of the substance as manufactured in the production scale. Entries in the IUCLID Section 1.2 Composition and Section 1.4 Analytical information will be based on the experience with lab trials.

A clear indication of how long ECHA will need to evaluate Inquiry dossiers is not given within the legal text, therefore concerning the time frame for this task is a great uncertainty for industry. It might be the bottleneck in the registration process for non-phase-in substances.

If a company intends to bring a new product or a new substance to market, it may be clever to wait as long as possible before the public and competitors gain this sensitive information. By submitting the inquiry dossier to ECHA competitors will be informed at an early stage about the intentions of the inquirer and maybe competitor to be. The consequence will be that the Lead Company or the first registrant may be in a position to abuse the situation easily. They are not allowed to hinder the inquirer in his intention to register, but it is highly likely that they can try to invoice costs at a high level from the potential registrant that will lead to further discussions that may lead to a delay in coming to market for the former inquiring party. Costs to be paid for the Joint submission will lead to a higher price for the product and therefore either marketing of the substance can be prevented or the general public has to pay for it.

Non-EU entities have the advantage that they can manufacture and bring a substance or product to market outside of the EU and also provide EU customers (each of them can purchase amounts below 1 t/a) without any one having the obligation to register under REACH. If the substance is marketable the Non-EU manufacturer will already have earned some money at the moment when he may make the decision to register this substance under REACH via an Only Representative. Furthermore, he will benefit from the fact that only amounts delivered to EU customers have to be considered for the registration under REACH. If a substance cannot be marketed successfully the Non-EU manufacturer did not lose any further money for registration purposes under REACH. This means Non-EU companies have a lower business risk.

This could lead to a reduction in innovation within the EU although REACH [1] intended to support innovation.

For globally acting enterprises it may bear the chance to develop new substances and new products outside of Europe to benefit from confidentiality and not having to spend money for the registration before the substance was brought to market successfully. As Switzerland is near the EEA it may be a great chance for Swiss companies to develop in this niche.

8.4
Examples and Exercises

1. Your company is located in Europe and intends to manufacture in the next year some substances that actually are not in the portfolio of your company. What needs to be checked previously to the decision that an Inquiry dossier has to be submitted to ECHA? Define which actions are necessary to ensure REACH compliance within your company concerning the above stated intention (describe the considerations and actions that may be necessary before making the decision that your company may have to submit an inquiry dossier).

2. Which IUCLID chapter or section has to be filled in for an Inquiry Dossier even if it may not be checked for certain types of registration dossiers later on? Why is this chapter of high importance for ECHA?

3. Your company intends to manufacture for the first time a non-phase-in substance. It shall be manufactured in an amount of 10 t/a. Production capacity will be 1050 kg/day. The first production campaign shall start on 1 June 2014. When is the registration deadline for your company? When do you have to start activities concerning REACH issues?

4. If you started too late with your activities concerning preparation of the registration of a certain non-phase-in substance that should be manufactured by your company (comprising of several sites within Germany, Belgium, Switzerland and India) what alternative solutions may be considered to be in the position to deliver the product in due time to your customer(s)?

References

1 Regulation (EC) No 1907/2006 of the European Parliament and of the Council of 18 December 2006, concerning the Registration, Evaluation, Authorisation and Restriction of Chemicals (REACH), establishing a European Chemicals Agency, amending Directive 1999/45/EC and repealing Council Regulation (EEC) No 793/93 and Commission Regulation (EC) No 1488/94 as well as Council Directive 76/769/EEC and Commission Directives 91/155/EEC, 93/67/EEC, 93/105/EC and 2000/21/EC.
2 see Article 3 (20) of the REACH regulation [1].
3 see Article 3 (20) (a) of the REACH regulation [1].
4 see Article 3 (20) (b) of the REACH regulation [1].
5 see Article 3 (20) (c) of the REACH regulation [1].
6 Notification of new chemical substances in accordance with Directive 67/548/EEC on the classification, packaging and labeling of dangerous substances, NO-LONGER POLYMER LIST, Version 3, published by the European Chemicals Bureau.
7 Guidance on registration, Version 2.0, May 2012, published by the European Chemicals Agency.
8 see Article 26 of the REACH regulation [1].
9 see Article 26 (3) of the REACH regulation [1].
10 see Article 26 (4) of the REACH regulation [1].
11 see Article 26 (2) of the REACH regulation [1].
12 see Article 26 (1) of the REACH regulation [1].
13 Guidance on data sharing, Version 2.0, April 2012, published by the European Chemicals Agency.

9
Process for Registration of Phase-In Substances

9.1
Preparing for Pre-registration and Late Pre-registration

REACH [1] entered into force on 01 June 2007. After that it lasted one year until it entered into operation on 01 June 2008.

From 01 June 2008 until 01 December 2008 each Manufacturer or Importer within the EU and every Non-EU manufacturer via Only Representative in the EU was allowed to pre-register phase-in substances [2]. Doing a pre-registration in general is free of charge and the pre-registrant is not obliged to register at a later stage, but will benefit from the extended registration deadline. Therefore, in case of doubt it was useful to pre-register a certain substance manufactured or imported by a company to avoid being left behind after the deadline for pre-registration.

A procedure for preparing an overview on the substances manufactured or imported by your company that needed to be pre-registered in due time previously to the relevant deadline in 2008 and that also may be used for first time manufacture or import until May 2017 for phase-in substances is described in Chapters 2 and 13 of this book including spreadsheets that may support you in defining the necessary actions.

In the case of first time manufacture or import it was possible and still is possible to do a late pre-registration after 01 December 2008 if there is left at least twelve months until the relevant registration deadline.

Companies that (late) pre-registered a substance can benefit from extended registration deadlines based on the relevant tonnage band and if appropriate the properties of a certain substance.

Registrations for phase-in substances could have been done starting with the entry into operation of REACH on 01 June 2008. For high volume chemicals manufactured or imported in amounts above 1000 t/a per manufacturer or importer and for volumes above 100 t/a in connection with properties that led to R50/53 and also for CMR substances from 1 t/a and above the registration deadline was 30 November 2010 if these substances were previously (late) pre-registered by a company.

Until 31 May 2013 all pre-registered substances manufactured or imported in the tonnage band 100 t/a up to 1000 t/a shall be registered.

REACH Compliance – The Great Challenge for Globally Acting Enterprises, First Edition.
Susanne Kamptmann.
© 2014 Wiley-VCH Verlag GmbH & Co. KGaA. Published 2014 by Wiley-VCH Verlag GmbH & Co. KGaA.

For the smaller tonnage bands 1 to 10 t/a and 10 t/a to 100 t/a the registration deadline may be 31 May 2018 if the corresponding substance had been (late) pre-registered in due time.

For an overview on timelines for pre-registration and registration see Figure 9.1).

Pre-registration is the process by which a potential registrant of a phase-in substance submitted the required information to ECHA, which consequently allowed him to benefit from the transitional regime for registration. The pre-registration period (1 June 2008–1 December 2008) is now over [3].

Late pre-registrations may still be done until 31 May 2017 for the smaller tonnage bands. This can be done online within only a few minutes in your company's REACH-IT account.

You will have to choose "pre-register substance online" from the menu "pre-registration" within your REACH-IT account. Then you will be required to agree to the following declaration [3]:

> "☐ I declare that the substance I pre-register is a phase-in substance according to Article 3 (20) of Regulation (EC) No 1907/2006 (REACH Regulation).
>
> I also declare that the substance I pre-register is **not**:
>
> (a) a phase-in substance classified as carcinogenic, mutagenic or toxic to reproduction, category 1 or 2, in accordance with Directive 67/548/EEC and manufactured in the Community or imported, in quantities reaching one tone or more per year;
>
> (b) a phase-in substance classified as very toxic to aquatic organisms that may cause long term adverse effects in the aquatic environment (R50/53) in accordance with Directive 67/548/EEC, and manufactured in the Community or imported in quantities reaching 100 tonnes or more per year;
>
> (c) A phase-in substance manufactured in the Community or imported, in quantities reaching 100 tonnes or more per year.
>
> ECHA and Member State Authorities reserve the right to verify the information provided. If the information provided is incorrect, you may be subject to enforcement actions by Authorities of the relevant Member States."

Afterwards you are asked for

- EC number/CAS number;
- tonnage band and relevant registration deadline;
- contact person within your company.

You may go on in the late pre-registration procedure by entering the requested information or choosing from a drop-down menu. After each relevant step you

9.1 Preparing for Pre-registration and Late Pre-registration

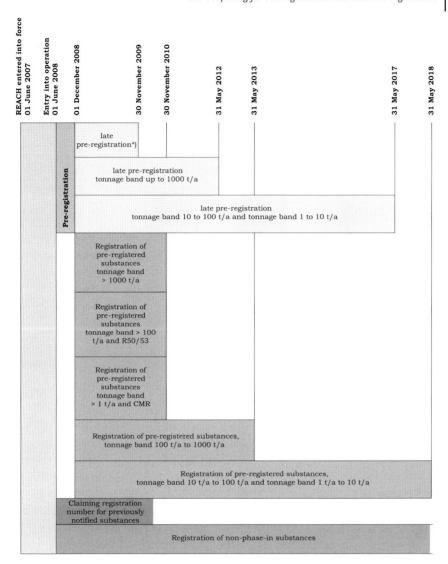

Figure 9.1 Timelines for pre-registration and registration of phase-in substances.
*) for all tonnage bands and all properties (phase-in substances).
Remark: Registration of pre-registered substances in the above scheme includes also registration of late pre-registered substances.

click on the "continue" or "next" button. Only if you entered all relevant information and it also seems appropriate to REACH-IT, will you be in a position to go forward.

After successful pre-registration previously to the 01 December 2008 a pre-registrant received a reference number (pre-registration number) starting with "05-", late pre-registrants receive a reference number starting with "17-" (see Figure 9.2).

Reference number for
- **pre-registration** previously to 01 December 2008:
  05-xxxxxxxxxx-xx-xxxx
- **late pre-registration** (from 01 December 2008 until 31 May 2017):
  17-xxxxxxxxxx-xx-xxxx

Figure 9.2 Format of a (late) pre-registration number.

After successful (late) pre-registration you will have access to the pre-SIEF page within REACH-IT. On this page you will find an overview on all companies that pre-registered a certain substance. You may see a certain pre-SIEF page only if your company pre-registered the same substance based on certain identifiers such as, for example, EINECS number or IUPAC name. The potential registrants among the pre-SIEF members may become SIEF members after agreement on the substance sameness based on a more detailed analysis and maybe in depth discussion.

The pre-registration number or late pre-registration number is only a proof for the registrant to be that he did a (late) pre-registration. The corresponding reference number needs to be included in the registration dossier later on. There is no need to provide your customers with the (late) pre-registration number.

Within the pre-SIEF page in REACH-IT there is no information whether a (late) pre-registrant acts as a manufacturer, importer or as an Only Representative. This information has to be given in the registration dossier later on.

Furthermore, on the pre-SIEF page in REACH-IT there will be no connection to the relevant tonnage bands or registration deadlines of other pre-SIEF participants. Tonnage band and relevant registration deadline that you entered by doing the (late) pre-registration for your own company can be checked at any time in REACH-IT. If necessary this information may be changed.

9.2
Communication within Pre-SIEF

Members of a pre-SIEF are listed in REACH-IT in connection to the same substance, but not all of them aim to register in future. There are three types of pre-SIEF members (see Figure 9.3). Data Holders, Third Party Representatives (TPR) and Potential Registrants (Manufacturers, Importers or Only Representatives acting on behalf of a Non-EU manufacturer) may be Members of a pre-SIEF.

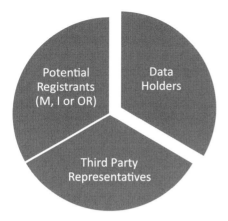

Figure 9.3 Different types of pre-SIEF members.

9.2.1
Data Holders

Within the group of data holders there may be data holders that automatically joined the SIEF and others that are acting on their own initiative.

Data holders that automatically join the SIEF may be

- any manufacturer or importer who has registered a phase-in substance before June 1, 2018 even without pre-registration (so-called early registrants) [4];
- parties that submitted information in the framework of the Plant Protection Product Directive (91/414/EC) or the Biocidal Product Directive (98/8/EC) [4].

Furthermore, there may be data holders that have the intention to share data that may be listed in REACH-IT on their own initiative. These data holders may be

- Manufacturers or importers of phase-in substances in quantities of <1 t/a who have not pre-registered [4].
- Downstream users of phase-in substances [4].
- Third parties holding information on phase-in substances:
  - Trade or industry associations, sector specific groups and consortia already formed [4].
  - Non-Governmental Organizations (NGOs), laboratories, universities, international or national agencies [4].
  - Manufacturers of a substance who have no interest in registering a substance under REACH, because they do not produce or place it on the market in Europe (e.g., a Non-EU manufacturer who does not export into the EU) [4].

9.2.2
Third Party Representatives

A Third Party Representative (TPR) can be appointed by manufacturers or importers of phase-in substances who have pre-registered these substances and producers or importers of articles who have pre-registered phase-in substances intended to be released from an article with the aim to hide their identity during eventual cooperation with other SIEF Members [4].

The TPR is then a SIEF member, but the legal entity that nominated the TPR retains the full responsibility for their REACH compliance [4]. However, the TPR will carry out all tasks concerning "Data Sharing and avoidance of unnecessary testing" [5].

9.2.3
Potential Registrants

Among the SIEF members normally the potential registrants are the subgroup with the highest number of companies.

Potential registrants can be

- current and potential manufacturers and/or importers of phase-in substances having pre-registered that substance [4];
- current and potential importers of preparations containing phase-in substances having pre-registered that substance [4];
- current and potential producers and importers of articles having pre-registered that phase-in substance if intended to be released from articles [4];
- Only Representatives of Non-EU manufacturers having pre-registered that phase-in substance [4];
- late pre-registrants (first time manufacture or import >1 t/a after 1 December 2008 [4, 6].

In REACH-IT there will be linked "A" (A = active) to all companies that are still willing to register a substance. If a manufacturer ceases production of a certain substance before the relevant registration deadline, he may inactivate the pre-registration that afterwards will have an "I" (I = inactive) linked to the corresponding company instead of the former "A". The same may apply if an importer pre-registered a certain substance to ensure legal compliance within the pre-registration period, but later on can benefit from being a downstream user as his suppliers all take care of the registration obligations (either in the role of a manufacturer/importer or a Non-EU manufacturer via an Only Representative).

9.2.4
Duties and Rights of the Different SIEF Participants

The different types or categories of SIEF participants may have different duties and rights within a certain SIEF.

List of duties and rights of potential registrants [4]	
Duties	**Rights**
1. React to requests from other participants	1. Request missing information from other SIEF participants
2. Provide other participants with "proof of cost" and (access to) existing studies upon request	2. Receive financial compensation for data shared
3. Agree on cost sharing mechanism	3. Submit joint registration (or decide to opt-out) Remark: Decision cannot be made by TPR but may be communicated in the SIEF by TPR
4. Collectively identify needs for further studies to comply with registration requirements	
5. Make arrangements to perform the identified studies	
6. Agree on classification and labeling	
7. Agree on the appointment of the Lead Registrant	
8. Agree on the selection of studies to be included in the joint submission	
9. Ensure that data sharing activities do not jeopardize compliance with EC competition law	

Figure 9.4 Duties and rights of potential registrants [4].

Data holders are only requested to react to any query from potential registrants, if they hold data relating to this query [7]. They do not have an active role in the SIEF [4].

The potential registrants and if so the TPRs appointed by potential registrants have the same duties within a SIEF (see Figure 9.4).

9.3
Formation of SIEF

To ensure that potential registrants of a pre-SIEF can work together efficiently there is somebody needed who coordinates the communication between the pre-SIEF members. Although there is no legal obligation to have a SIEF Formation Facilitator (SFF), from a pragmatic point of view it is useful that one company of the pre-SIEF is willing to take over the role of the SFF and indicates this in REACH-IT.

If your company pre-registered a great number of substances you may be interested in going ahead at first with the substances that need to be registered at an earlier date, because of their volumes relevant for your company or maybe because of their special properties that may have an influence on the registration deadline.

After having prioritized which substances your company would be willing to do the SFF tasks and eventually would also volunteer to become Lead registrant, you are recommended to indicate the role SFF in REACH-IT. Instead of the former "A" indicating that your pre-registration is active and your company intends to register a certain substance, then there will be a "F", indicating that your company is acting as the SIEF Formation Facilitator. Other companies will within REACH-IT for a certain substance find their company in the first row and in the second row the company acting as a SIEF Formation Facilitator.

Perhaps you will also add a short message in the "information field" within REACH-IT to inform pre-SIEF members about the next steps that you will undergo with the aim to fulfill your duties as a SFF.

If another pre-registrant has already selected the position SFF in the pre-SIEF page in REACH-IT, this option is no longer available for any other pre-SIEF member [4]. If the original SFF reviews its position, any other pre-SIEF participant on its way to registration may take over the role of the SFF.

As the first task to be fulfilled by the SFF will be to establish whether the potential registrants plan to register the same substance for the purpose of data sharing and harmonization of Classification & Labeling, the SIEF initially enables a discussion to establish sameness and organizes the exchange of information and data on the identity of the substance if necessary. Only when this is agreed is a SIEF created [4]. Former pre-SIEF members consequently will be SIEF members. In many cases where the pre-SIEF members pre-registered a certain substance based on the corresponding EINECS number this will be a task that easily can be fulfilled. More complicated may be cases in which multi-constituent substances or UVCBs are involved. In such cases pre-SIEFs may be split or eventually merged after in depth discussions within the pre-SIEF(s).

The necessary communication may be done by email in the beginning. Depending on the size of a SIEF it may be necessary especially for SIEFs comprising of a large number of companies to implement later a special IT-platform for the SIEF communication. Such an IT-platform will support the SFF and at a later stage the Lead Company in fulfilling their tasks concerning information exchange within the SIEF. Tasks to be done by a SFF are listed in Figure 9.5.

As the role of the SFF may be time consuming and may require substantial resources, it can be expected that SIEF members will financially compensate the SFF for their services. Apart from operational SIEF costs, the decision is up to the member of the SIEF [4]. Therefore, the role SIEF Formation Facilitator is not to be held in high esteem, but if your company intends to become Lead Company and wants to benefit from having an influence on the tasks that will be done and also on the costs that will arise on the way to preparation of a registration dossier, it may be acceptable to take the risk that your company will not receive compensation for the tasks done in the role of the SFF.

Tasks to be done by a SFF	
Task	**Forms and documents to be prepared to fulfill a certain task**
• Facilitate discussion between potential registrants on the sameness of the substance [4]	• Take pre-SIEF information from REACH-IT as an xml-File and transfer it into an Excel-Sheet • Prepare a Substance Identification Profile (SIP) • Document when you sent out emails (when, to whom, content and corresponding responses)
• Find the form of cooperation and internal rules for the SIEF [4]	• Pre-SIEF Survey for determination of SIEF Category (Lead, involved, passive dormant) • Survey "Lead registrant Agreement"
• Launch the query for data in SIEF [5]	• Pre-SIEF Survey
• Prepare an inventory of available data [4]	• Data gap analysis
• Perform the technical work	
• Channel the cooperation with other SIEFs and data holders [4]	
• Ensure a smooth entry of late pre-registrants [4]	
• Facilitate an agreement on cost sharing [4]	

Figure 9.5 List of tasks that may be done by a SFF.

The SFF may initiate several pre-SIEF surveys concerning different topics such as SIEF status, substance sameness, providing a SIP, asking for data and studies available within pre-SIEF members' companies and finally asking (pre-)SIEF members concerning their agreement on appointing the Lead Company.

In cases where a substance is identified as a mono-constituent substance with a CAS number and EINECS number available and the number of pre-SIEF members is not too high, it seems possible to cover several of these tasks within only one single pre-SIEF survey.

The easiest way to reach all relevant pre-SIEF members will be to export the information on pre-SIEF members including their email addresses from REACH-IT as an xml-file. This xml-file can easily be converted into an Excel-file just by opening it in Excel. The relevant email addresses then may be copied quickly and entered into the field for recipients in the email you intend to sent to the pre-SIEF members.

Example letters for the pre-SIEF survey can be found in Figures 9.6 and 9.9. Such a letter may either be attached to the email that the SFF sends to all potential registrants or the questions may be included directly into the email.

If your company has more capacities or you deal with a great number of pre-SIEF members you may have implemented an IT-platform to do the pre-SIEF survey(s). In this case you may only be required to inform all pre-SIEF members once with access information concerning this IT-platform.

> **REACH: Lead Registrant Agreement concerning [substance name]**
>
> Dear pre-SIEF member,
>
> Our company [company name] is a member of the pre-SIEF for the substance
> **[substance name], CAS [CAS number], EINECS [EINECS number]**
> for which your company has also submitted a (late) pre-registration.
> We are writing this letter in the role of the SIEF Formation Facilitator (SFF).
>
> The aim of this letter is to ask you whether it would be convenient for your company that [company name] takes over the role of the Lead Registrant of this SIEF. If you do not reply until [dd/mm/yyyy], we assume that you agree with [company name] becoming the Lead Company.
>
> Furthermore, we would appreciate that you indicate your SIEF category – if you do not reply to this survey until the above stated deadline, we will assume your answer to be "4 – dormant".
> Assignment of SIEF codes is not static and you may ask the SIEF Facilitator to change it at any time.
>
> Overview of SIEF categories:
> 1 (Leading). This is a substance for which my company has available resources to (co-)lead and drive registration to completion.
> 2 (Involved). My company has the intention to register and may be actively involved.
> 3 (Passive). My company has the intention to register this substance and is interested to receive information about progress in the SIEF.
> 4 (Dormant). My company has no intention to register. For this category there is no need to answer the survey as status 4 will be automatically assigned to the companies that do not respond.
>
> May we also ask whether you have data relevant for the registration of this substance?
>
> We are looking forward to your reply.
>
> Best regards,
> [company name]

Figure 9.6 Example letter for a pre-SIEF Survey including Lead Registrant Agreement.

9.3.1
Substance Sameness and Substance Identification Profile (SIP)

In order to reach an agreement on the sameness of a substance, potential registrants must enter into pre-SIEF discussions. As a consequence, a SIEF is formed when the potential registrants of a substance in the pre-registration list agree that they effectively manufacture, intend to manufacture or import a substance that is sufficiently similar to allow a valid joint submission of data [8].

The SFF, or if already elected the lead registrant, may provide a substance identification profile (SIP) to all other potential registrants as a starting point for eventual necessary discussions. An example of a simple Substance Identification Profile can be found in Figure 9.7. Templates for more detailed SIPs were published by CEFIC [9].

Substance Identification Profile (SIP)

Substance name:
Synonyms:
CAS number: [….]
EC number: [….]

Molecular formula:
Structural formula:

Molecular weight or molecular weight range:

Degree of purity: ☐ monoconstituent substance, purity > 80% - ≤ 100%

Impurities: No impurity relevant for classification

Short description of the analytical methods:

Figure 9.7 Example for a simple Substance Identification Profile.

Relevant criteria for substance identification can be found in the corresponding Guidance document [10].

All sameness discussions shall be conducted in compliance with EC competition law while protecting confidential business information. Therefore, the SFF may prepare a substance identification profile that is not too detailed, at least in the beginning. For a mono-constituent substance it may be enough to state that the main constituent identified by CAS number and EINECS number shall have a concentration above 80% and there should not be included any impurity in a concentration that would require a change of classification. Then it will not be disclosed which manufacturing process a particular company follows. If it cannot be avoided to differentiate and it seems to be necessary to exchange more in depth or detailed information for the purpose of evaluation of sameness it may be necessary to appoint an independently acting Trustee for example, a consultant company to avoid any action that would mean a breach of EC competition law. The same may be necessary for multi-constituent substances or UVCBs.

9.3.2
Lead Registrant Agreement

Under the REACH regulation [1] the role of the Lead Registrant is a mandatory role laid down in Article 11 (1) [11].

The Lead registrant is defined as the "one registrant acting with the agreement of the other assenting registrant(s)" and it is he who must first submit certain information [8].

REACH does not specify rules as to how the Lead registrant should be selected [8]. However, the lead registrant must act with the agreement of the other potential

> **Who shall act as the Lead Registrant?**
>
> • If only one potential registrant volunteers to become lead registrant he needs to persuade the other potential registrants to agree to appoint him as lead registrant [8];
>
> • If two or more potential registrants volunteer to become lead registrant, they can seek an agreement between themselves as to who will be the lead registrant and request endorsement by all potential registrants. If the volunteers cannot agree, then it is recommended that the other potential registrants appoint the lead registrant [8].
>
> • If no potential registrant volunteers to become lead registrant, the lead registrant may be the EU manufacturer or EU importer with the highest interest in registration (e.g. highest tonnage, most data, etc.). However, the lead registrant still needs to be endorsed by all potential registrants [8].

Figure 9.8 Options to define who shall act as the Lead Registrant.

registrants. There may be several options to define or determine who shall act as the lead registrant for a certain substance (see Figure 9.8).

As registrants to be often have the experience that it will be difficult to have the written approval from all potential registrants that they agree on the company volunteering to become the lead registrant, because very often there will be no response from pre-registrants that do not intend to register any longer or maybe at a later stage, it is highly recommended to do the request for approval in an appropriate way.

It was efficient when a "lead company to be" asked for their appointment among the pre-SIEF members by asking for approval by assuming that no response will mean approval. No response can also be connected to a SIEF status 4, that means "dormant". For an example questionnaire of a pre-SIEF Survey including the Lead Registrant Agreement, please see Figure 9.9. If it seems more appropriate you may do several surveys within the pre-SIEF and use only parts of the example questionnaire in each of the surveys. Whether you first prepare a letter on headed paper of your company that may be attached to an email or rather write directly an email is up to you. Even when your company prefers doing the survey via an IT-platform you may use some of the questions presented in the example letter.

In any case it is important that you document all your actions with due diligence to avoid future problems. Document exactly when you sent out any information and to whom. In case an email-address of any pre-SIEF member that you found in REACH-IT is no longer valid and your information or email concerning a survey cannot be delivered to the intended recipient, please make a note of this. All responses to your survey shall be recorded accordingly.

To ensure that late pre-registrants can contact you at any time, it is recommended not only to indicate in REACH-IT that your company is SIEF Formation Facilitator, but also to add information at the moment you are an elected Lead Registrant. In any case you should actualize your information in REACH-It whenever it seems necessary, especially when the contact person within your company changes.

Please return to:
[company name, address and contact details]

Lead Registrant Agreement

CAS number	
EC number	
Substance name	

1. Do you accept [company name] as Lead Registrant of the SIEF for [substance name]?

   ☐ Yes, we accept [company name] as Lead Registrant
   ☐ No, we do not accept [company name] as Lead Registrant

2. SIEF category

   ☐ Leading (SIEF Code 1)
   ☐ Involved (SIEF Code 2)
   ☐ Passive (SIEF Code 3)
   ☐ Dormant (SIEF Code 4)

3. Do you have data relevant for the registration of this substance?
   ☐ Yes
   ☐ No

   If yes, are these data involving tests on vertebrate animals?
   ☐ Yes
   ☐ No

4. Should there be a consortium associated with this substance, would you be interested in joining?
   ☐ Yes
   ☐ No
   ☐ Don´t know

5. Are you interested in buying a LoA? – If so, please indicate the relevant year.
   ☐ 2010
   ☐ 2013
   ☐ 2018

_____   _____   _____
Company name   date, place   signature

Figure 9.9 Example questionnaire for a pre-SIEF Survey including Lead Registrant Agreement.

The lead registrant may be one of those registrants who plan to submit their registration dossier by the earliest registration deadline from among all the potential registrants [8], but that is not a must.

However, the lead registrant has to submit a dossier well before the registration deadline of the company having the earliest deadline within the SIEF and the registration dossier shall contain data in accordance with the highest data requirements within the SIEF. The lead registrant itself will have to pay the fee with

ECHA only corresponding to his own tonnage, even when there are assenting registrants having a higher tonnage band.

9.3.3
Lead Registrant Notification

ECHA encourages Lead Registrants and Candidate Lead Registrants to identify themselves via the Lead Registrant notification. This can be done within only a few minutes on ECHA's homepage on the site "Lead Registrant notification". ECHA distinguishes between Lead Registrant and Candidate Lead Registrant.

A Candidate Lead Registrant may be a company who is interested in taking up the role as Lead Registrant but has not been nominated yet within the SIEF. This role does not assume any legal obligations, but may help ECHA to target specific issues in SIEF management.

As soon as your company has officially been appointed as Lead Registrant, for example, after having done the corresponding survey in the pre-SIEF, you may notify your company as the Lead Registrant.

ECHA publishes on their homepage a list of "Active lead registrants". The lead registrant name will be disclosed, if the lead registrant has allowed ECHA to do so. As a consequence, the lead registrant can no longer claim it confidential in the registration dossier. ECHA also checks the new substance registrations on a regularly basis and updates the list of active Lead registrants to indicate when a dossier has been submitted by a lead registrant. There will also be a hint when a dossier was submitted by a different (lead) registrant (who did not notify ECHA).

For the online notification as a Lead Registrant you shall provide the status of the substance for the lead company and furthermore provide EC/list number, CAS number and IUPAC name or other chemical name. Furthermore, ECHA asks for information concerning the earliest registration deadline by which the Lead Company intends to submit the joint registration, the estimated number of SIEF members willing to register and the highest tonnage band of the joint submission. Concerning the elected Lead Registrant the company UUID is required (can be found in your REACH-IT Account) and some details concerning the contact person within the Lead Company.

9.4
Cooperation within the SIEF

The functioning of the SIEF, to be agreed by all SIEF participants, may be detailed in a SIEF agreement. SIEF participants are free to choose the form and the clauses to be included in such an agreement. The agreement may consist of a combination of SIEF operating rules, participation processes, data and cost sharing mechanisms and other important aspects that the SIEF participants may consider on a case by case basis [8]. Different types of standards and templates of agreements are already available and used by different industries for data sharing purposes [8].

9.4.1
Obligations of SIEF Participants

All SIEF participants must agree to the appointment of a lead registrant according to Article 11 (1) [11].

Furthermore, all SIEF members as potential registrants shall (see also Figure 9.4):

- React to requests for information from other participants (within one month according to Article 30 (1) [12].

- Provide other participants with existing studies both those on vertebrate animals and others, if requested [8].

- Request missing data information related to vertebrate animal testing from other SIEF participants; they may also request other non-animal data from other SIEF participants [8].

- Collectively identify needs for further studies to comply with registration requirements [8].

- Make arrangements to perform the identified studies [8].

- Agree on classification and labeling where there is a difference in the classification and labeling of the substance between potential registrants [8]. However, there may be more than one classification and labeling, in a given joint registration dossier (e.g., because of different impurities) [8].

9.5
Data Sharing

The members of the SIEF are free to choose how to organize their cooperation under REACH. The form of cooperation can vary from a simple structure (e.g., IT tools to communicate between all SIEF members) to a more structured and complex organization (e.g., legally established consortia). For large SIEFs, consortia may be a more efficient type of cooperation to provide a binding means of complying with the data sharing obligations and to prepare the registrations. However, there is no requirement to form consortia under the REACH regulation [13].

In any case, it is important to keep in mind that REACH is not a competition-law-free zone and therefore in any cooperation all aspects of the competition law have to be respected.

If there shall be a consortium established from a large SIEF it may be necessary to have an independently acting trustee (e.g., consultant company) for doing the communication within the consortium. The trustee may also calculate the costs to be invoiced from each member based on the individual tonnage band or data requirements depending on the contractual situation. Furthermore, the trustee may evaluate studies and tests that are available from members of the consortium and calculate the value of these studies to define the price that has to be shared by the members of the consortium. It is clear that a consultant will have hourly rates that will lead to invoices that have to be paid by the consortium. As each member will have to pay its share, it should be in the interest of all members to keep things as simple as possible and to avoid unnecessary work concerning the SIEF management to be done by the consultant on behalf of the members.

Within a smaller group of potential registrants it may be possible to avoid costs for a trustee doing the SIEF management or consortium management, because the Lead Company is in the position to do these tasks on their own by respecting the competition law.

As a consortium in any case will require larger efforts for doing all necessary tasks, because there are involved so many companies that all have the right to discuss all issues previously to finalization of the registration dossier of the lead company and this will be time consuming and also cause higher costs, it should also be considered whether a SIEF can agree on the Lead company acting as the Lead registrant on behalf of the other SIEF members, but can do all necessary decisions on their own. Later, the Lead Company could sell a Letter of access to all assenting registrants. There is a great chance that the time consuming decision-making processes that are required within a consortium can be avoided and costs can be kept at a lower level. On the other, hand the members of Joint submission to be will not have too much influence on the content of the registration dossier and if so the Chemical Safety Report or the costs spent by the Lead company for any task.

Therefore, it is very often time consuming to find a solution on how to cooperate within a certain SIEF. The type of cooperation may be different for each substance as in each SIEF there are different companies having different strategies on how to cope with their obligations under REACH. It is clear that every type of agreement or other contract needs to be checked by the parties willing to sign and therefore it may be expected that large numbers of lawyers will be busy with providing and checking of all the different contracts. Later, they may have further work in all cases of disagreement concerning these contracts for example, when parts of a contract seem to be not acceptable because of breach of any regulation or there are discrepancies in connection with other valid documents.

In any case the involvement of lawyers will lead to further costs to be paid by all registrants to be.

Whenever possible it is recommended to use at least as a basis for any contractual arrangement within the SIEF a model that already has been checked and either can be used directly or needs just smaller amendments to be adapted to the needs of a certain group of SIEF members. CEFIC offers on their homepage [14] some models for agreements in the SIEF that are proven to be neutral and therefore are already widely used.

REACH defines the obligations of (pre-)SIEF members as data sharing, agreeing on classification and labeling and also doing a joint submission, but there are no rules or hints on how to do it. However, it seems to be important to have a framework for the SIEF process and SIEF management including details on organization of dossier preparation, cost sharing criteria, liabilities and how each potential registrant may participate in the joint submission.

It is not possible to deal with all possible contractual arrangements in detail within this book, but we will have a short glance at the well-known models among the agreements in the next sections.

9.5.1
Consortium Agreement

The comprehensive consortium agreement is recommended for SIEFs with a high number of members as the work to be done is organized with legal certainty [15].

9.5.2
Cooperation Agreement

The cooperation agreement may be used for smaller SIEFs and when there are several companies willing to be involved and therefore will be interested in signing a cooperation agreement. The cooperation agreement normally will be signed only by the leadership team within a SIEF. The preparation of the registration dossier including all data requirements that have to be sent by the lead company on behalf of all the other members of joint submission will be done mainly by the leadership team. Additionally, the Lead Registrant or the leadership team may offer a SIEF Agreement to all other registrants to be. A Model Cooperation Agreement was published by CEFIC [16].

9.5.3
SIEF Agreement

A SIEF Agreement may be offered by a Lead Registrant or a leadership team to all SIEF members that intend to register. It is a sort of standardized legal framework that will cover SIEF operating rules, data sharing rules and also cost sharing criteria. The obligations and liabilities of the lead registrant are clearly defined. Cost sharing may be explained in a more general way, as there are in the beginning no exact figures available. Therefore, a SIEF Agreement will only set up basic rules or principles.

9.5.4
Letter of Access

A Letter of Access (LoA) may be appropriate in different situations. Well known are the Letter of Access used to grant a registrant or a group of registrants the right to refer to studies and other data from a third party, but there is also a contractual arrangement possible that allows a registrant to be to become a member of joint submission and refer to a registration dossier of a lead company as a whole.

9.5.4.1 Letter of Access Concerning Data as Studies and Tests

A letter of access granting the registrant(s) the right to refer to a certain study of a Data Holder may either grant this right to only one single registrant or it will grant transferable rights for example, when a lead company buys this LoA on behalf of all registrant(s) to be for a certain substance.

If a Data Holder gives at first a right to refer to a certain study only to one registrant for example, the lead registrant for a certain substance, it will be required that further registrants for the same substance later on will have to negotiate with the Data Holder, too. In principle, it will be necessary that each registrant to be buys his own LoA from this data holder previous to becoming Member of Joint submission and consequently previous to submitting their registration dossier. This may be extremely time consuming and also open the doors to a grey zone, as the lead company has to give a hint to all registrants to be that they will have to buy a LoA for a certain study from the corresponding Data Holder. Afterwards, every registrant to be should in principle negotiate with the Data Holder. The latter will have to offer a LoA to all parties requesting this, by considering the number of requesting parties in total to grant a fair cost sharing. This means after selling a further LoA the Data Holder should calculate whether the former buyers of such a LoA shall be reimbursed. The administrative burden for doing such a procedure in a fair and transparent way normally may not be justifiable, although the Data Holder may be interested in knowing who will use his study and this time-consuming procedure is the only one that enables the Data Holder to know each registrant who will refer to his study.

Another problem may be that the lead company does not know whether a registrant to be already bought a LoA from a certain Data Holder and it is not really its obligation to check this, therefore the situation may occur that the lead company may provide a registrant to be with submission name and token and enable a member of joint submission to register without having fulfilled their obligations towards the Data Holder. There will be almost no chance for the Data Holder to find out that a certain registrant as a Member of Joint submission refers to his study as it was used by the Lead Company and therefore it may be difficult to claim a share from this party. The situation will be absolutely not transparent in cases were the study of the Data Holder is required for higher tonnage bands and therefore may not be required by all registrants that are members of joint submission. The Data Holder in such a case has to trust on the support of the lead company and the honesty of the registrants that will lead to a situation where the Data Holder, contrary to his intention, does not have control over which of the members of a joint submission refers to his study, if the lead company does not inform him accordingly.

To avoid such difficulties and the really time-consuming administrative tasks by granting the right to refer to each party on their own, it is recommended to Data Holders to provide the Lead Company with a LoA to a certain study by granting transferable rights to the Lead Company for example, restricted to REACH registrations concerning all registrants that intend to register the same substance. Cost sharing for the LoA to this study then has to be managed by the Lead Registrant, but not by the Data Holder and it can be ensured by the Lead Company that only parties that shared the costs for this study may become a member of joint submission.

9.5.4.2 Letter of Access to a Registration Dossier

A registrant to be may not have any studies on his own and maybe he is not in the position to spent a lot of time for tasks to be done in a consortium or he is not interested in being involved in the whole process for dossier preparation, then it may be appropriate for this "registrant to be" to buy a Letter of access that will give him the permission to refer to the registration dossier of the Lead Company. Normally a company that buys a LoA from the Lead company receives neither a copy of the registration dossier nor copies of the studies or robust study summaries that were used to prepare such a dossier, but they will receive submission name and token from the Lead company that give them the possibility to become a Member of Joint submission previous to the submission of their own dossier.

This type of Agreement is the most efficient way to save time for the company purchasing the LoA. As it also avoids a lot of time spent by the Lead company for preparing a comprehensive contract as it is used for consortia and there will be no in depth discussion and no meetings necessary the SIEF management will not bind too much resources within the Lead Company. Hence, this sort of Agreement normally leads to a low price for the LoA. Very often the Lead dossier is already accepted by ECHA and overall costs are known by the Lead company, therefore the companies buying a LoA will often benefit from having clear figures concerning the costs. Very often the Lead company has a fixed price for the purchasers of LoAs and no further costs are to be expected. If the price for a single LoA is low, because of a high number of registrants buying such an LoA, the administrative efforts for a refund in the case of selling further LoAs will be higher than the refund itself, therefore a Lead company and also the buyers of an LoA may agree on a fixed price for the LoA.

However, if there are only a few companies willing to share costs previous to a certain registration deadline and it is still open as to what will happen in future it may be necessary to have some more details included in a LoA document. It may be considered that ECHA could have future demands concerning studies and tests that will cause high costs to the Lead company but is also for the benefit of the co-registrants. Therefore, the Lead company may intend to determine that future costs have to be shared also by parties that purchase a LoA. On the other hand, the party willing to buy a LoA may insist on having a refunding if the costs are actually shared only between a few companies, but there may be a higher number of registrants to be that may buy a LoA at a later stage.

Although the LoA contract in general is the simplest version of offering access to a registration dossier of the Lead company, the overall procedure including countersigning of the contract, issuing the invoice and payment that has to be done by the party interested in buying the LoA may last several months. In case you are interested in buying a LoA from a Lead company, please ask for it in due time and to be on the safe side calculate a time period of 6 months up to 8 months until you will receive submission name and token.

9.6
Data Sharing Disputes

In Article 30 [7] of the REACH regulation [1] the European Parliament tried to determine rules for data sharing within the SIEF. The first section in Article 30 (1) [12] says: "Before testing is carried out in order to meet the information requirements for the purposes of registration, a SIEF participant shall inquire whether a relevant study is available by communicating within his SIEF. If a relevant study involving tests on vertebrate animals is available within the SIEF, a participant of that SIEF shall request that study. If a relevant study not involving tests on vertebrate animals is available within the SIEF, a SIEF participant may request that study." Furthermore, it is described in this Article [12] that the owner of the requested study has the obligation to answer within one month to such a request by providing a proof of its cost. The participants shall make every effort to ensure that the costs of sharing the information are determined in a fair, transparent and non discriminatory way [12]. Already this point may be subject of arguments, as very often the situation may occur that a study owner either did a study or test within his own company and therefore does not have any invoice or the invoice for a certain study or test was lost for example, because of splitting a company or when the study was done within a group of companies that participated somehow in the costs. To avoid ongoing problems because of missing invoices or in case the company requesting a certain study has the feeling that the price paid originally by the owner may have been too high, there had been an approach to standardize the value of studies and tests. Very often the so-called "Fleischer list" [17] is cited in the procedure to determine the value of a certain study. This list is based on average prices that had to be paid in the past for a certain study. Insiders claim that on top of the prices given in the Fleischer list should be added additional costs to cover the inflation rate and, furthermore, there should be considered also the costs for the analytical examinations that have to be done in addition to a certain study for example, to determine the concentration of a substance used for this type of study. When considering the measurement of a certain sample the time spent for development of the analytical method also has to be considered. This can mean that there are further costs of several thousand € that may have to be added to the study value itself to fix the costs that have to be compensated within the SIEF. As a data owner had to bear the risk in the case of difficulties that could have arisen and maybe he had to pay for repeating a study when the results were not clear or the study was not exactly executed as foreseen because of any problem that occurred at the first trial, it also seems appropriate to have a sort of advantage compensation from the requesting party that will benefit from not having this sort of risk any longer. It must be assumed that all these details will lead to ongoing discussions and arguments in daily life especially in cases where direct competitors are obliged to deal with each other to fulfill the demands of REACH by respecting the competition law at the same time. The statement that cost sharing in a fair, transparent and non discriminatory way shall be done is a nice sentence, but as experience shows it can be difficult and time consuming to find an agreement. However, at the moment an agreement has been reached and payment is done by the requesting party, Article 30(1) [12] lays down that the owner shall give permission to refer to the full study report for the purpose of registration within two weeks. The last sentence within Article 30 (1) [12] clarifies: "Registrants are only required to share in the costs of information that they are required to submit to satisfy their registration requirements." This may again lead to further trouble within SIEFs. If a co-registrant intends to register a certain substance as a transported isolated intermediate in a tonnage band below 1000 t/a he is in general not urged to provide any studies or tests as detailed in Annex VII to Annex X [18] and therefore will not have to share the costs for that. On the other hand, the Lead registrant may have the same intention concerning the type of registration and tonnage band but feels obliged to fulfill the demands of CLP in the classification part of the registration dossier and therefore uses some studies or tests. Therefore, it seems not to be fair when only the lead company shall have the costs for such studies, whereas the co-registrants would benefit from that part of the dossier without any cost sharing.

A similar situation may occur when ECHA demands further studies or tests from the Lead company for example, as a result of an evaluation process. First, the Lead company will have to take further action to fulfill the demands of ECHA in due time, but furthermore has the risk that members of joint submission will not support the lead company by taking their share immediately, although this is foreseen in accordance with Article 53 (2) [19]. In such a case Article 30 (2) [20] demands that all reasonable steps shall be taken to reach an agreement within the SIEF, but it will be difficult for the Lead company to wait until an agreement has been reached. Very often timelines will demand that they start with further actions to have the study results in due time previous to having the Agreement within the SIEF. Maybe afterwards Article 53 (4) [21] has to be applied where it is determined that "The person performing and submitting the study shall have a claim against the others accordingly. Any person concerned shall be able to make a claim in order to prohibit another person from manufacturing, importing or placing the substance on the market if that other person either fails to pay his share of the cost or to provide security for that amount or fails to hand over a copy of the full study report of the study performed. All claims shall be enforceable in the national courts. Any person may choose to submit their claims for remuneration to an arbitration board and accept the arbitration order" [21]. However, if it should be necessary to fight for a justified cost compensation there will always be additional costs for a lawyer and maybe the court previous to having the situation clarified. Hopefully, SIEF members will be able to find agreements that can be accepted by all parties without involving lawyers and courts with disputes.

9.7
Opt-Out

Opt-out is a topic that under REACH is discussed often, but is not absolutely clarified yet. In principle, ECHA urges all members of a SIEF that intend to register the same substance to do that jointly. This means that all registrants to be that intend to register a certain substance shall be member of a joint submission.

The Lead Company will submit their registration dossier first and afterwards all members of joint submission may submit their dossiers. This procedure seems to be clearly structured, but reality shows a different picture.

Previous to the first registration deadline it was difficult to provide the dossiers in time as timelines were really tight, whereas fulfilling data requirements was connected to the high volume tonnage bands more challenging than it is for smaller tonnage bands with less data requirements and longer periods till the registration deadline in 2013 or 2018. Therefore, very often registrants to be in 2009 and 2010 were worried, because they did not know whether the elected Lead Company would meet the timelines in such a way that members of joint submission would also be able to register in due time. Nobody was really experienced in submitting dossiers to ECHA and no registrant to be could be sure that ECHA would accept their dossier after first submission.

Therefore, in cases where a company intended to register a substance having a tonnage band of only 100 to 1000 t/a in connection with R50/53 as a transported isolated intermediate, they often made the decision to submit their dossier as a single submission. The higher registration fee seemed to be negligible compared to an interruption of manufacture or import of a certain substance that would have been the consequence if the substance was not registered in due time.

Very often a company that had only one or at least only few competitors had the fear that they could fail in meeting the deadlines, whereas a competitor acting as the Lead Company could have done the registration just in time for themselves, but not so early that all other members of joint submission could have met the same deadline. In particular, the risk of failing in the first submission of a dossier made a lot of companies think about alternatives to the joint submission. Under the given circumstances a company that should have registered as a member of joint submission could have registered a certain substance previously to the elected Lead registrant. By doing so, it was also possible for the company that did the separate registration to act as a Lead Registrant. As it is the SIEF members' obligation but not ECHA's to agree on the Lead Registrant, it may be assumed that ECHA will not be in the position to penalize companies that did separate registrations in the past.

ECHA so far did nothing against this practice and it still works. If a Lead company registered a substance with standard requirements, whereas another company registered only a transported isolated intermediate or an on-site isolated intermediate it may be even accepted by ECHA that a SIEF was split.

It will also be justified to split a SIEF if members of a SIEF find out that they do not have the same substance for example, because of a different impurity profile of the substance to be registered that leads to different study results and different classifications.

However, in general SIEF members shall do a joint submission based on the principle "one substance one registration". Justifications for an Opt-out are described in Article 11 (3) [22] and in Article 19 (2) [23].

Article 11 (3) [22] may be applicable for standard registrations as it allows that "A registrant may submit the information referred to in Article 10 (a)(iv), (vi), (vii) or (ix) separately if:

a) it would be disproportionately costly for him to submit this information jointly; or

b) submitting the information jointly would lead to disclosure of information which he considers to be commercially sensitive and is likely to cause him substantial commercial detriment; or

c) he disagrees with the lead registrant on the selection of this information." [22]

As information referred to in Article 10 (a)(iv), (vi), (vii) or (ix) concerns only parts of the dossier, ECHA still expects that all registrants are a member of joint submission. If a company wishes to do an opt-out concerning parts of the dossier, it is required that this company enters certain data that normally are provided by the Lead Company on behalf of all other registrants, in their IUCLID substance data set. By doing so they indicate that they want to do an opt-out concerning this information. ECHA will always expect that a registrant that does an opt-out, because Article 11 (3) (a), (b), or (c) are applicable, gives an "explanation as to why the costs would be disproportionate, why disclosure of information was likely to lead to substantial commercial detriment or the nature of the disagreement, as the case may be" [22]. A registrant doing an opt-out, furthermore, has to take the risk that ECHA will check their dossier in depth especially the parts that are relevant for the opt-out. If data or information in this part seems not to be appropriate, ECHA may reject the dossier or alternatively may ask for further information.

For the registration of intermediates there are also some cases known in which an opt-out may be justified. In Article 19 (2) [23] it is defined similar to Article 11 (3) [22]: "A manufacturer or importer may submit the information referred to in Article 17 (2) (c) or (d) and Article 18 (2) (c) or (d) separately if:

a) it would be disproportionately costly for him to submit this jointly; or

b) submitting the information jointly would lead to disclosure of information which he considers to be commercially sensitive and is likely to cause him substantial commercial detriment; or

c) he disagrees with the lead registrant on the selection of this information." [23]

The consequences are the same as in the case of an opt-out that is done for a standard registration. The registrant has to justify for the opt-out and risks a more in depth check of his dossier by ECHA. A short summary of the key messages concerning opt-out is given in Figure 9.10.

> - REACH allows registrants to opt-out only from a part of the joint submission under the circumstances defined in Article 11 (3) [22] and Article 19 (2) [23]
>
> - Companies will have to pay fees for a single submission even when a dossier contains only one single opt-out option
>
> - Registrants must comply with their REACH obligations by proceeding as per Article 30 of the REACH Regulation (i.e. data sharing). Registrants who opt-out must still participate in the joint submission [8]
>
> - Registrants are allowed to opt-out for certain or all given endpoints but must remain member of the joint submission [8]
>
> - Note that even when the registrant decides to exercise his opt-out option, he remains a member of the joint submission and will be able to submit his dossier only after the lead dossier has been accepted for processing. Hence, a registrant can opt-out from certain information requirements but not from the joint submission as such. [24]

Figure 9.10 Key messages concerning Opt-out.

9.8
Registration Dossier of the Lead Company and Registration Dossiers of the Members of Joint Submission

A Lead Company will have to prepare a dossier including data and information as required for the company having the highest data requirements within the Joint submission even if the Lead company itself will have fewer data requirements. Each member of the Joint submission may submit their own registration dossier but not before the one of the Lead company has passed the enforce rules check and therefore is accepted for further processing.

Members of Joint submission will have to provide only a few chapters as they can refer to the parts of the dossier of the Lead company that were submitted on behalf of all members of the Joint submission. If necessary, the Chemical safety Report in full or in parts may be submitted by the Lead Company on behalf of the members of Joint submission, but it is also possible that each registrant does it on its own. In the case of a transported isolated intermediate each registrant has to provide the Guidance on safe use in Chapter 11 of the IUCLID Substance data Set on his own, for standard registrations this chapter may be done by the Lead Company on behalf of the other members. For in depth information concerning the preparation of dossiers either for the Lead Company or for a Member of Joint submission, please go back to Chapter 6 of this book.

Keep in mind that filling in chapters or sections that are normally not required in a member dossier will be considered as an opt-out concerning the information provided in this chapter.

9.9
Overview on Important Steps within the Process for Registration of Phase-In Substances

As a short summary for the important steps within the REACH registration procedure for phase-in substances the scheme below (Figure 9.11) was compiled. It is simplified as late pre-registration is not considered and the type of cooperation within the SIEF is not determined.

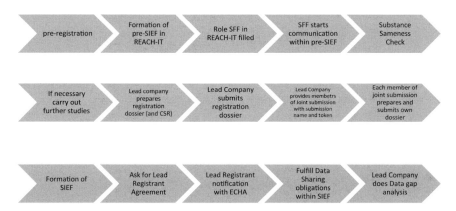

Figure 9.11 Overview on important steps in the process for registration of phase-in substances.

9.10
Examples and Exercises

1. What types of pre-SIEF members may be found in REACH-IT for a certain substance?

2. Assume that your company located in London (United Kingdom) intends to manufacture a certain phase-in substance for the first time in 2015. Will it be possible to do a late pre-registration? Give a justification for your answer.

3. Is there a difference between a Candidate Lead Registrant and an elected Lead Registrant?

4. Does an Opt-out mean that a registrant to be may deviate from the "one substance one registration principle"?

5. When will a member of joint submission be allowed to submit their own dossier at the earliest?

References

1 Regulation (EC) No 1907/2006 of the European Parliament and of the Council of 18 December 2006, concerning the Registration, Evaluation, Authorisation and Restriction of Chemicals (REACH), establishing a European Chemicals Agency, amending Directive 1999/45/EC and repealing Council Regulation (EEC) No 793/93 and Commission Regulation (EC) No 1488/94 as well as Council Directive 76/769/EEC and Commission Directives 91/155/EEC, 93/67/EEC, 93/105/EC and 2000/21/EC.
2 See Article 3 (20) of the REACH regulation [1].
3 REACH-IT Industry User Manual, Part 04 – Late pre-registration, Version 2.0, July 2012, published by ECHA.
4 REACH Industry Preparation Letter No 9, June 2008, published by CEFIC.
5 See Title III, Article 25 to 30 of the REACH regulation [1].
6 See Article 28 (6) of the REACH regulation [1].
7 See Article 30 of the REACH regulation [1].
8 Guidance on data sharing, Version 2.0, April 2012, published by ECHA.
9 http://www.cefic.org/Documents/IndustrySupport/SIP_template_final.xls.
10 Guidance for identification and naming of substances under REACH and CLP, Version 1.1, November 2011, published by ECHA.
11 See Article 11 (1) of the REACH regulation [1].
12 See Article 30 (1) of the REACH regulation [1].
13 REACH-IT Industry User Manual, Part 05 – Pre SIEF, Version 2.0, July 2012, published by the European Chemicals Agency.
14 www.cefic.org (accessed September 2013).
15 Navez, V. (2012) Agreements in the SIEF, LR Workshop, ECHA, 03 February 2012.
16 http://www.cefic.org/Documents/IndustrySupport/Cefic-REACH-cooperation-agreement-between-SIEF-Lead-Members_21.09.09_final.doc.
17 Fleischer, M. (2007) "Research paper: testing costs and testing capacity according to the REACH requirements – Results of a Survey of Independent and Corporate GLP Laboratories in the EU and Switzerland." *Journal of Business Chemistry*, **4** (3), 96–114.
18 See Annex VII to Annex X of the REACH regulation [1].
19 See Article 53 (2) of the REACH regulation [1].
20 See Article 30 (2) of the REACH regulation [1].
21 See Article 53 (4) of the REACH regulation [1].
22 See Article 11 (3) of the REACH regulation [1].
23 See Article 19 (2) of the REACH regulation [1].
24 Guidance on registration, Version 2.0, May 2012, published by ECHA.

10
What Happens after Submission of Your Registration Dossier to ECHA?

Independent of whether you submit a PPORD notification, an Inquiry Dossier or a registration dossier under REACH [1] you will create the requested dossier type within your IUCLID5 Account. It is highly recommended that you check the dossier by using the Technical-Completeness-Check-Plug-in and also the Fee-calculation tool before exporting and submitting the dossier. The exported .i5z File has to be stored in your computer and afterwards may be imported into REACH-IT. IUCLID5 and REACH-IT are two separate IT-tools on purpose. In the IUCLID5 Account the registrant may work on Substance Data Sets and create dossiers without ECHA being able to check anything in your account. The IUCLID 5 Account "belongs" to industry, whereas any action that is done in REACH-IT will be done "under the eyes" of ECHA.

Before you can do any submission, you will need an Account in REACH-IT for your company. The dossier that shall be submitted to ECHA must be in the format .i5z and must be uploaded into REACH-IT. In this chapter you will find a short description of what happens next to your successful upload. Parts of the further process steps are done automatically, some may also be done by ECHA staff manually.

There will be a slightly different procedure for updates of registration dossiers compared to the initial registration.

10.1
Initial Verification

All dossiers submitted to ECHA undergo a number of initial technical and administrative checks in order to ensure that they can be handled properly and that the required regulatory processes can be successfully carried out. The different initial checks (see Figure 10.1) are described below in the chronological order in which they take place [2].

In the first step of the initial verification every dossier is scanned for known viruses [2]. Depending on the outcome of this scan the dossier is either rejected

| Upload of dossier in REACH-IT | **Initial verification – step 1** Virus Scan | **Initial verification – step 2** File format validation | **Initial verification – step 3** Internal structure validation | **Initial verification – step 4** Business rule validation | Dossier accepted for further processing or rejection |

Figure 10.1 Initial verification.

> **Submission number:**
> <2 uppercase letters><6 digits>-<2 digits>
> e.g.: **WX005880-13**

Figure 10.2 Format of a submission number.

or will proceed to the next step (if there are no viruses found). Step 2 checks whether the potential registrant uploaded an .i5z File, all other formats will lead to immediately rejection. In step 3 all attachments are checked concerning their format. For Chemical Safety Reports it is foreseen that registrants attach this document as a .pdf to Chapter 13 of the dossier. Other formats for example, a Chemical Safety Report attached as a Word document in the format .doc or .docx may lead to a message that is sent to you in your REACH-IT message box, but the dossier can be further processed or the dossier may be rejected for example, because of attachments that are in a format that is not supported or recognized by REACH-IT [2]. The technical completeness-Check-Plug-in (TCC Plug-in) in IUCLID 5 is not able to recognize this, therefore every registrant is recommended to check this manually before submitting any dossier.

The REACH-IT software checks in step 4 whether the business rules are fulfilled. The business rules are a set of prerequisites that must be fulfilled before ECHA can establish that the dossier can be accepted for further processing [2]. As only some of the business rules can be checked by using the TCC Plug-in in IUCLID 5 it is highly recommended to check other critical points manually with due diligence before a dossier is submitted via REACH-IT. The dossier may be accepted for further processing only if all of the relevant business rules are satisfied [2]. If this is the case you will receive this good news in a message from ECHA sent to your REACH-IT message box. In the case of rejection you will be informed accordingly. In the case of rejection at this stage REACH-IT will not remember that you already tried to submit a dossier for a certain substance although there will be a line in the menu registration with the remark "failed". This means that in the case of rejection you should find out where you failed, afterwards correct this in the Substance Data Set within your IUCLID 5 Account or in the Dossier Header when creating a new dossier. The new dossier may be submitted to ECHA and the above stated procedure will start again.

After successful business rule validation a submission number is generated for each submission by REACH-IT. The submission number (see Figure 10.2) is a unique number issued for every submission but does not provide any information regarding the dossier type or company information.

The format is: 2 uppercase letters, 6 digits, a hyphen and two digits [3]. The last two digits are used as checksum [3]. The submission number will be used in all further communications from REACH-IT to the submitting party concerning a certain dossier. The submission number of the latest successful submission concerning a certain substance will be needed for the preparation of a future update.

10.2
Overall Completeness Check

The Overall Completeness Check consists of two Checks that are both performed in REACH-IT. These two Checks are the Technical completeness check and the Financial completeness check [2]. ECHA normally will undertake the completeness check of a registration dossier within three weeks of the submission date. For registration dossiers of previously pre-registered substances that are submitted within the two-month period immediately preceding the registration deadline ECHA will have the duty to undertake this completeness check within three months [4]. Therefore, it must be recommended to submit registration dossiers of pre-registered substances more than 2 months before the relevant deadline to ensure that registration is finalized in due time.

The Technical completeness check is performed for all types of registration including their updates and for PPORD notifications. Details concerning the chapters and sections of the dossier that are checked can be found in Chapter 6 of this book. In principle, this technical completeness check can be done by each registrant by using the TCC Plug-in in IUCLID 5.

If there is no failure so far REACH-IT will open an invoice (if applicable) and the registrant is asked to make the payment within 14 days. Fees will be invoiced based on type of submission and for a registration dossier the type of registration by considering the size of the registrant's company and also whether it is a single submission or a joint submission. Details concerning the fees to be expected can either be found in the "fees order" [5] or calculated by using the fee calculation Plug-in in IUCLID 5.

It is recommended to pay in due time, although there will be just a friendly reminder if you do not manage to pay in due time and the due date will be extended. It is experienced that it will take about 5 to 7 days from the transfer of the money until ECHA find the money in their account. In the meantime, the procedure within REACH-IT does not proceed. After ECHA having received the payment in REACH-IT the Financial completeness Check will be "succeeded" and the Overall Completeness Check will be started.

If a registration is incomplete, ECHA shall inform the registrant, before expiry of the three-week or three-month period referred to in the second subparagraph of Article 20 (2) [4], as to what further information is required in order for the registration to be complete, while setting a reasonable deadline for this [4]. The registrant shall complete his registration and submit it to ECHA within the deadline set [4]. This completed registration dossier must be identified as an update [2] and the submission number of the previous submission has to be indicated in the dossier header. ECHA will perform a further completeness check, considering the further information submitted [4].

ECHA shall reject the registration if the registrant fails to complete his registration within the deadline set. The registration fee shall not be reimbursed in such cases [4].

If there is no indication to the contrary from ECHA within the three-week or three-month period of the submission date the registrant may start or, in the case of a phase-in substance that had been (late) pre-registered, continue without interruption the manufacture or import of the substance [2].

10.3
Receiving the Reference Number

When Technical completeness Check and Financial completeness Check both are passed the reference number (in the case of registration dossiers known as registration number) will be issued as defined in Article 20 (3) [6]. A reference number has been coined as a more general designation than registration number or notification number and represents a unique number that is generated by REACH-IT and given to a substance and a company after dossiers of certain types are successfully submitted for the first time [3].

For different types of dossiers the submitting party will receive different reference numbers. The structure of a reference number is

<TYPE>-<BASE NUMBER>-<CHECKSUM>-<INDEX NUMBER> [3],

where <TYPE> is a 2-digit number, the <BASE NUMBER> is a 10-digit number generated randomly, the <CHECKSUM> is a 2-digit checksum and <INDEX NUMBER> is a 4-digit number that indicates the index of a Member of Joint submission [3].

An overview on different types of reference numbers is given in Figure 10.3.

In the case of a registration the registration date will be the same as the submission date [6]. The registrant will have the duty to use the registration number (or reference number) for any subsequent correspondence regarding registration procedures [2, 6].

10.4
End of Pipeline Activities

There are some further duties for ECHA. In accordance with Article 20 (4) [9] within 30 days of the submission date, ECHA has to notify the Competent Authority of the Member State within which the manufacture takes place or the

Type of dossier submitted to ECHA	Type of reference number	
Registration	Registration number [7]	01-xxxxxxxxxx-xx-xxxx
C & L notification	C & L notification number	02-xxxxxxxxxx-xx-xxxx
Substance in article	Substance in articles notification number	03-xxxxxxxxxx-xx-xxxx
PPORD	PPORD notification number [8]	04-xxxxxxxxxx-xx-xxxx
Pre-registration (done before 30. November 2008)	Pre-registration number	05-xxxxxxxxxx-xx-xxxx
Inquiry	Inquiry number	06-xxxxxxxxxx-xx-xxxx
Data Holder notification		09-xxxxxxxxxx-xx-xxxx
Downstream User notification	Downstream user report number	10-xxxxxxxxxx-xx-xxxx
Application for Authorization		11-xxxxxxxxxx-xx-xxxx
Substance Evaluation		12-xxxxxxxxxx-xx-xxxx
Annex XV – C & L Harmonization		13-xxxxxxxxxx-xx-xxxx
Annex XV - Authorization		14-xxxxxxxxxx-xx-xxxx
Annex XV - Restriction		15-xxxxxxxxxx-xx-xxxx
Internal usage		16-xxxxxxxxxx-xx-xxxx
Late pre-registration (done after 30. November 2008)		17-xxxxxxxxxx-xx-xxxx

Figure 10.3 Reference numbers released for certain types of dossiers.

importer is established that the registration submitted [2]. ECHA will inform the Competent Authority about the information that ECHA stored in their database. The information comprises the registration dossier including submission number and reference number, submission or registration date, result of the completeness check and any request for further information and corresponding deadline set [9]. If the manufacturer has production sites in more than one Member State, the relevant Member State shall be the one in which the head office of the manufacturer is established [9]. The other Member States where the production sites are established shall also be notified [9]. The Agency shall forthwith notify the competent authority of the relevant Member State(s) when any further information submitted by the registrant is available on the Agency database [9]. This means Competent Authorities of the Member States will be informed after submission of any Update.

After finalization of all necessary end of pipeline activities the registrant or party that submitted a dossier for any other purpose to ECHA will be informed accordingly via the message box in REACH-IT. In the case of registrations ECHA will provide the registrant with a decision letter in which it is stated that the registrant has the permission to manufacture or import a certain substance in the tonnage band that was given in the registration dossier.

Although the submission process or registration process is finalized for the time being, the registrant cannot expect that ECHA in future will no longer check the dossier.

10.5
Dossier and Substance Evaluation

Dossier evaluation [10] and Substance Evaluation [11] are subject of Title VI [12] in the REACH regulation [1].

10.5.1
Examination of Testing Proposals

First, ECHA will examine any testing proposal set out in a registration or a downstream user report [13]. Priority shall be given to registrations of substances that have or may have PBT, vPvB, sensitizing and/or carcinogenic, mutagenic or toxic for reproduction (CMR) properties, or substances classified as dangerous according to Directive 67/548/EEC above 100 tonnes per year with uses resulting in widespread and diffuse exposure [13]. Information relating to testing proposals involving tests on vertebrate animals shall be published on the ECHA website [14]. Within the following public consultation lasting 45 days from the date of publication, anybody is invited to submit scientifically valid information and studies of relevance [14]. Subsequently, ECHA will prepare a draft decision as laid down in Article 40 (3) [15] based on the procedure laid down in Article 50 and Article 51 [16]. ECHA may decide that the registrant has to carry out the proposed test and will also set a deadline for submission of the study summary, or the robust study summary. ECHA may in this decision ask the registrant to modify the conditions under which the test is to be carried out and/or demanding that the registrant shall do one or more additional tests. Furthermore, ECHA may reject the testing proposal or if several registrants submitted proposals for the same test concerning a certain substance they may decide to give several registrants the chance to reach an agreement on who will perform the test on behalf of all of them [15]. The registrant in any case has the chance to sent a comment to ECHA concerning the draft decision within 30 days after having received the draft decision [17]. If the registrant submitted a comment to ECHA, then ECHA has to inform the competent authority without delay. The competent authority and ECHA will take any comments received into account as described in Article 50 [17] and they may amend the draft decision accordingly. Cooperation of ECHA and the competent authorities of the Member States is further defined in Article 51 [18]. The procedure is the following: After ECHA having notified its draft decision and the comments of the registrant to the competent authorities of the Member States, the Member States may propose amendments to the draft decision to ECHA within 30 days. If ECHA does not receive any proposals for amending the draft decision, the former draft decision will become a final decision. If ECHA modifies the draft decision based on

10.5 Dossier and Substance Evaluation

Time period for Preparation of draft decision by ECHA	
Non-phase-in substances	**Phase-in substances**
Preparation of draft decision within 180 days of submission date [20]	By 1 December 2012 for all registrations received by 01 December 2010 containing proposals for testing (Annex IX and Annex X) [21]
	By 1 June 2016 for all registrations received by 01 June 2013 containing proposals for testing (Annex IX only) [21]
	By 1 June 2022 for any registrations received by 01 June 2018 containing proposals for testing [21]

Figure 10.4 Time periods for examination of testing proposals and preparation of a draft decision by ECHA.

proposed amendments from the Member States, there will be a 15-day period, beginning after the deadline of the 30-day period in which Member States were allowed to sent proposals for amendment to ECHA, in which ECHA may submit the draft decision and the amended version of it to the Member State Committee. Any registrant has again 30 days time to comment, whereas the Member State Committee shall reach an agreement on the draft decision within 60 days. When the Member State Committee reaches a unanimous agreement, ECHA will amend the former draft decision into a final decision. If the Member State Committee does not reach a unanimous agreement the Commission will have to prepare a draft decision with the procedure laid down in Article 133 (3) [19].

Time periods for examination of testing proposals are different for non-phase-in substances compared to phase-in substances. The draft decision shall for non-phase-in substances be prepared by ECHA within 180 days of receiving a registration or downstream user report containing a testing proposal [20]. For phase-in substances ECHA prepares in accordance with Article 40 (3) [15] the draft decisions until the dates given in Article 43 (2) [21]. A short summary of the deadlines until when ECHA shall submit a draft decision concerning examination of testing proposals is given in Figure 10.4.

After having received a final decision the registrant is obliged to take further action accordingly and provide ECHA with an updated dossier in due time.

10.5.2
Compliance Check of Registration

In accordance with Article 41 [22] ECHA may examine any registration at any time to verify that the information in the technical dossier(s) submitted complies with the requirements and that any adaptions of the standard information requirements and the related justifications comply with the rules governing such adaptations [22]. Furthermore, chemical safety assessment and a chemical safety report may be checked concerning their compliance with the requirements [22]. On the basis of such an examination ECHA may, within 12 months of the start of the compliance check, prepare a draft decision requiring the registrant to bring the registration(s) into compliance [22]. In any case, the list of dossiers being checked for compliance by ECHA shall be made available to Member States competent authorities [23].

To ensure that registration dossiers comply with the REACH regulation [1], ECHA shall select a percentage of the dossiers, no lower than 5% of the total received for each tonnage band, for compliance checking [24]. ECHA will give priority to dossiers containing information provided by registrants that were doing an Opt-out, but ECHA will also give priority to certain parts of the dossier(s) or do random tests.

During 2012 ECHA examined almost every dossier for transported isolated intermediates in Section 3.5 to check whether the entered uses were in accordance with the conditions given in Article 18 (4) [25] that have to be fulfilled if a registrant intends to benefit from reduced data requirements for this type of registration.

If ECHA does not have any objection concerning the examined part of the dossier, the registrant will not be informed. But maybe at a later stage the same dossier will be checked again in part or in full. Therefore, a registrant cannot be sure at any time that he does not have to provide further information in future.

10.5.3
Substance Evaluation

The objective of substance evaluation is to clarify an initial concern for human health or the environment [26]. ECHA shall, in cooperation with the Member States, develop criteria for prioritizing substances with a view to further evaluation [27]. Prioritization shall be on a risk-based approach [27]. Hazard information, exposure information and aggregated tonnage from the registrations submitted by several registrants shall be considered [27]. Based on these criteria ECHA is compiling a draft Community rolling action plan (Corap) which shall cover a period of three years and shall specify substances to be evaluated each year [28]. Substances may be included based on dossier evaluation carried out by ECHA or based on any other source for example, notification from a Member State to ECHA [29] if there is a reason for the assumption that a certain substance constitutes a risk to human health or the environment [28]. ECHA will adopt the final Community rolling action plan in cooperation with the Member State Committee. The Member State who will carry out the evaluation of a certain substance is identified in the Corap. ECHA coordinates the substance evaluation process, but the competent authorities of Member States will carry out the evaluation [30]. The competent authorities of the relevant Member State may appoint another body to act on their behalf [30]. Although a compliance check is not mandatory for substance evaluation, substances listed in the Community rolling action plan should be given priority for compliance check because it will facilitate the substance evaluation [26].

The substance evaluation process provides a mechanism for competent authorities of Member States, where necessary to require (the) registrant(s) to obtain and submit additional information to address the initial concern [26, 31]. Following substance evaluation, the Member State Competent Authority may come to the conclusion that action should be taken under the authorization, restriction or classification and labeling procedures in REACH, or that no further action is needed [26].

More detailed information on the Authorization process and restrictions will be given in Chapter 12 of this book.

10.5.4
On-site Isolated Intermediates are not the Object of Evaluation

For on-site isolated intermediates that are used under strictly controlled conditions neither dossier nor substance evaluation shall apply [32].

However, if the competent authority of the Member State in whose territory the site is located considers that a risk to human health or the environment, equivalent to the level of concern arising from the use of substances meeting the criteria for inclusion in Annex XIV [33], arises from the use of an on-site isolated intermediate and that risk is not properly controlled, may take further action [32]. The competent authority of the Member State may require the registrant to submit further information directly related to the risk identified [32]. Such a request shall be accompanied by a written justification [32]. The competent authority of the Member State may also recommend any appropriate risk reduction measures [32]. If a competent authority does such an evaluation, they shall inform ECHA of the results and afterwards ECHA will inform the competent authorities of the other Member States [32].

10.6
Further Obligations of the Registrant and Downstream Users

After having successfully submitted a registration dossier for a certain substance a registrant and sometimes also the Downstream Users have some further duties that can bear some more challenges on the way to achieving REACH compliance.

10.6.1
Safety Data Sheets and extended Safety Data Sheets

A supplier of a substance or a preparation shall provide the recipient of the substance or preparation with a safety data sheet compiled in accordance with Annex II [34] in the following cases [35]:

- substance or preparation meets the criteria for classification as dangerous in accordance with Directives 67/548/EEC or 1999/45/EC; or [36]
- substance is persistent, bioaccumulative and toxic or very persistent and very bioaccumulative in accordance with the criteria set out in Annex XIII [37]; or [38]

- substance is included in the list established in accordance with Article 59 (1) for reasons other than those referred to above [39].

When supplying a substance on its own, the SDS has to be prepared for the substance itself, when supplying a substance in a mixture, the SDS has to be prepared for the mixture [2].

Annex II of the REACH regulation [34] defines the requirements for Safety Data Sheets. The allowed uses (in accordance with information given in the registration dossier) have to be mentioned. For a standard registration where a CSR was submitted with the registration dossier exposure scenarios must be considered in the Safety Data Sheet. The final Exposure Scenario (ES) developed for identified uses as part of the Chemical Safety Assessment (CSA) has to be communicated to the registrant's customers as an annex to the Safety Data Sheet [2]. Such a Safety Data Sheet is known as an extended Safety Data Sheet or eSDS. Manufacturers or importers located within the EU have to register the whole amount of a substance manufactured or imported above 1 t/a and therefore they will have to provide all their customers with the corresponding registration number included in the Safety Data Sheet. If a customer intends to sell as a distributor such a substance again to his customers he will have to prepare his own Safety data Sheet by omitting the last four digits of the registration number.

More difficult is the situation for Non-EU manufacturers that registered a substance via an Only Representative located within the EU. In this case the customers within the EU have to be provided with Safety Data Sheets including name and address of the Only Representative and also the complete registration number. If the Non-EU manufacturer also purchases Non-EU customers it can be recommended to provide those customers with a Safety Data Sheet that does not contain a registration number, as these amounts in general must not be registered and therefore very often were not considered in the registration dossier. Problems could occur if the Non-EU customers were provided with the registration number and will at a later stage forward the substance to a European country without having informed the Only Representative of the Non-EU manufacturer that registered this substance. In particular, if the registration had been done in a tonnage band below 1000 t/a there may occur a situation where in a certain calendar year a certain amount was manufactured and parts were delivered to a Non-EU customer that stored this amount for a while. If the Non-EU customer has the idea to bring this amount into the EU in a later calendar year and the registrant itself will bring again a certain amount of the substance into the EU directly both amounts may be summed up without the registrant/Only representative knowing that. If the tonnage band increases because of this, the authorities may expect the registrant to do an update of the registration dossier, whereas the Non-EU customer forwarding the substance bought at an earlier stage might not even know that there is a problem at all. To avoid such an issue it is highly recommended that Non-EU manufacturers that registered via an Only Representative provide only direct EU customers with the registration number.

If a Non-EU customer desires to bring the substance bought from a Non-EU manufacturer to a site within Europe either the importer in Europe has to register on his own or the Non-EU customer may ask his supplier whether he is willing to cover these amounts with his registration to support his Non-EU customer in his further business. If the Non-EU manufacturer that registered via OR agrees that the amounts sent by his Non-EU customer to Europe shall be recorded, then the Only Representative shall be informed accordingly.

It is the responsibility of the supplier to keep the SDS updated [2].

10.6.2
Documentation of Correspondence with Customers Purchasing Transported Isolated Intermediates

If your company registered a substance as a transported isolated intermediate that previously had been pre-registered it was possible to supply all customers with the pre-registered substance until the submission of the registration dossier independently of their uses. However, normally you will have asked your customers concerning their uses or they may have provided the registrant with their uses to make them identified uses to ensure that they are Downstream Users without any doubt.

At the latest after having registered a substance as a transported isolated intermediate you have to insist on your customers providing you with a confirmation that they use the substance for the synthesis of another substance in accordance with REACH Article 18 (4) [25]. See also Chapter 5 of this book.

10.6.3
Substance Volume Tracking

After having registered a certain substance in a certain tonnage band it is required to do a check on a regularly basis whether the chosen tonnage band is still in accordance with the actual amounts manufactured or imported by the registrant. ECHA expects registrants to inform them without undue delay of changes such as tonnage band increase or cessation of manufacture. A registrant located within the EU has to consider the whole amount manufactured or imported in each calendar year. Therefore, Substance Volume Tracking can be easily done by using the usually available IT-tools. There will be only exceptional cases if a substance can be regarded as registered under REACH but amounts are exempted because they fall under the scope of another regulation, for example, use of a substance in medicinal products.

For a Non-EU manufacturer that registered via an Only Representative the Substance Volume Tracking may be more difficult as he may not use the amounts manufactured within a calendar year, but has to consider the amounts that were brought to the EU within a certain calendar year. When there are amounts of the substance delivered to Non-EU traders or Non-EU manufacturers that may deliver parts of the substance at a later stage to Europe the calculation of the relevant tonnage can become really difficult and time consuming. Actually, there is no standard IT-tool available that can record this without further expensive adaptations. Therefore, normally the Substance Volume Tracking will be done manually by Non-EU manufacturers that registered via an Only Representative to be in the position to provide the Only Representative with correct figures. These figures may be requested from national authorities for on-site inspections or inspections at the Only Representative. As there is a strong cooperation between customs and national authorities it is highly recommended to take this task seriously.

10.6.4
Obligation to Update Information

Article 22 [40] defines that "following registration, a registrant shall be responsible on his own initiative for updating his registration without undue delay with relevant new information and submitting it to ECHA" in the cases listed in Article 22 (1) (a) to (i) [41]. We will see that in detail in Chapter 11 of this book.

As the wording "without undue delay" is not further defined and it is also not defined how a registrant shall receive knowledge of any new information, industry has to find a specific procedure concerning this task.

Consultants often recommend to do a literature search on a regular basis once or twice per year, but especially for very well known phase-in substances it may be considered to do this not that often as time and costs for the search may not be justified in correlation to the new information that may be found only seldom for these substances. This may be different for cases where within a consortium or a SIEF it was decided to do further examinations and maybe it was agreed on doing further tests for example, concerning the classification and labeling, then the outcome of the tests may be included in a dossier as soon as possible and an update should be submitted to ECHA at least by the Lead Company. When the information affects also parts of the dossiers that are submitted by each Member of a Joint submission on its own, then also the Members of Joint submission shall consider submitting an update.

10.6.5
Obligations of Downstream Users

Downstream Users do not have the obligation to register on their own, but they are obliged to cooperate with their suppliers/the registrants. They will have to consider whether they can agree on signing an article 18(4) confirmation when their supplier offers only a transported isolated intermediate instead of doing a standard registration.

After having received a Safety Data Sheet including a registration number they will have to check whether their uses are covered. In the case of eSDS this may be time consuming but will be absolutely necessary to find out whether further action will be required. If one or several of his uses are not considered in the registrant's dossier the Downstream User may ask his supplier to do an update that considers these uses or alternatively may submit a Downstream User Report to ECHA [42].

10.7 Examples and Exercises

1. If your company uploaded a registration dossier in REACH-IT that failed because of a failure in Business Rules you may intend to do an update. Do you have to provide the submission number of the first trial?

2. What is the difference between a reference number and a registration number?

3. What information can be extracted from the following reference numbers?

Reference number	Information
01-xxxxxxxxxx-28-0000	
01-xxxxxxxxxx-28-0001	
05-xxxxxxxxxx-xx-0000	
02-xxxxxxxxxx-xx-0000	

4. Which format has a reference number received by a company after having submitted an inquiry dossier?

5. What is the difference in the format of any reference number and a submission number?

6. Please check the following combinations of numbers and letters carefully. Which may be a valid submission number or a reference number? If it is neither a submission number nor a reference number, please give a justification for that.

"Combination of numbers and letters"	Submission number	Reference number	Neither submission number nor reference number – justification
LE900070-68			
MAX899605-98			
Pz899643-85			
01-611273564-78-0002			

7. For an update of a dossier where has the registration number and the submission number of the previous submission to be entered in IUCLID?

8. How can you find out what fees are applicable for a certain type of dossier that you intend to submit to ECHA for the first time?

9. Your company submitted a registration dossier for a previously pre-registered substance manufactured in amounts of 700 t/a at your site to ECHA on 28 March 2013. Assume the dossier passed the enforce rules check and was accepted for further processing, when will you have a response from ECHA at the latest? Can the same be applied if the dossier was submitted on 01 April 2013? Could there occur any problem for the continuous production of this substance at your site?

References

1 Regulation (EC) No 1907/2006 of the European Parliament and of the Council of 18 December 2006, concerning the Registration, Evaluation, Authorisation and Restriction of Chemicals (REACH), establishing a European Chemicals Agency, amending Directive 1999/45/EC and repealing Council Regulation (EEC) No 793/93 and Commission Regulation (EC) No 1488/94 as well as Council Directive 76/769/EEC and Commission Directives 91/155/EEC, 93/67/EEC, 93/105/EC and 2000/21/EC.

2 Guidance on registration, Version 2.0, May 2012, published by the European Chemicals Agency.

3 REACH-IT Industry User Manual, Part 06 – Dossier submission, Version 2.0, July 2012, published by the European Chemicals Agency.

4 See Article 20 (2) of the REACH regulation [1].

5 COMMISSION IMPLEMENTING REGULATION (EU) No 254/2013 of 20 March 2013 amending Regulation (EC) No 340/2008 on the fees and charges payable to the European Chemicals Agency pursuant to Regulation (EC) No 1907/2006 of the European Parliament and of the Council on the Registration, Evaluation, Authorisation and Restriction of Chemicals (REACH).

6 See Article 20 (3) of the REACH regulation [1].

7 In accordance with Article 20 (3) of the REACH regulation [1].

8 In accordance with Article 9 (3) of the REACH regulation [1].

9 See Article 20 (4) of the REACH regulation [1].

10 See Article 40 to Article 43 of the REACH regulation [1].

11 See Article 44 to Article 48 of the REACH regulation [1].

12 Title VI comprises of Articles 40 to Article 54 of the REACH regulation [1].

13 See Article 40 (1) of the REACH regulation [1].

14 See Article 40 (2) of the REACH regulation [1].

15 See Article 40 (3) of the REACH regulation [1].

16 See Article 50 ("Registrants' and downstream users' rights") and Article 51 ("Adoption of decisions under dossier evaluation") of the REACH regulation [1].

17 See Article 50 (1) of the REACH regulation [1].

18 See Article 51 ("Adoption of decisions under dossier evaluation") of the REACH regulation [1].

19 See Article 133 (3) of the REACH regulation [1].

20 See Article 43 (1) of the REACH regulation [1].

21 See Article 43 (2) of the REACH regulation [1].

22 See Article 41 ("Compliance check of registrations") of the REACH regulation [1].

23 See Article 41 (2) of the REACH regulation [1].

24 See Article 41 (5) of the REACH regulation [1].
25 See Article 18 (4) of the REACH regulation [1].
26 Guidance on Dossier and Substance Evaluation, June 2007, published by the European Chemicals Agency.
27 See Article 44 (1) of the REACH regulation [1].
28 See Article 44 (2) of the REACH regulation [1].
29 See Article 45 (5) of the REACH regulation [1].
30 See Article 45 of the REACH regulation [1].
31 See Article 46 (1) of the REACH regulation [1].
32 See Article 49 ("Further information on onsite isolated intermediates") of the REACH regulation [1].
33 See criteria laid down in Article 57 ("Substances to be included in Annex XIV") of the REACH regulation [1].
34 See Annex II ("Guide to the compilation of Safety Data Sheets") of the REACH regulation [1].
35 See Article 31 (1) of the REACH regulation [1].
36 See Article 31 (1) (a) of the REACH regulation [1].
37 See Annex XIII of the REACH regulation [1].
38 See Article 31 (1) (b) of the REACH regulation [1].
39 See Article 31 (1) (c) of the REACH regulation [1].
40 See Article 22 ("Further duties of registrants") of the REACH regulation.
41 See Article 22 (1) of the REACH regulation [1].
42 Guidance for downstream users, January 2008, published by the European Chemicals Agency.

11
Update of the Registration Dossier

11.1
When to Update Your Registration Dossier

There are in general two reasons possible why registrants have to update their registration dossiers: either it is an "update on request" as a consequence of a decision made by ECHA or it is a "spontaneous update" (see Figure 11.1).

There is a duty to keep information submitted to ECHA in a registration dossier up to date [1, 2]. It is the responsibility of the registrant to update his registration information when needed [3]. If the update of a dossier is submitted on a registrant's own initiative it falls into the category "spontaneous update".

If the information to be updated is part of jointly submitted information, it will be the lead registrant who will have to update the registration on behalf of the members of the joint submission [3]. Depending on whether the information to be updated concerns only information that was provided by a single registrant on his own or that has to be submitted individually by each registrant (e.g., Chapter 11 "Guidance on safe use" for registrants of transported isolated intermediates) it may be the case that concerning a certain substance only one registrant, several members of a joint submission or all members of a joint submission will have to submit an updated dossier to ECHA.

The registrant will have to update his IUCLID5 dossier and submit it to ECHA through REACH-IT [3]. Where the update relates exclusively to administrative data such as the identity of the registrant or the composition of the group of registrants in a joint submission, however, the updated information will be directly reported in REACH-IT [3]. No update of the IUCLID5 dossier is required in this case [3].

Reasons for an update on the registrant's own initiative are given in REACH Article 22 (1) [2]. We will treat this in depth in Section 11.3 of this book.

Whenever a registrant is requested to prepare an update of a dossier in IUCLID5 he will have to indicate the reason for the submission of this update in the dossier header. Either "update on request" or "spontaneous update".

For a requested update there must be provided also the communication number from ECHA's request in the dossier header.

REACH Compliance – The Great Challenge for Globally Acting Enterprises, First Edition.
Susanne Kamptmann.
© 2014 Wiley-VCH Verlag GmbH & Co. KGaA. Published 2014 by Wiley-VCH Verlag GmbH & Co. KGaA.

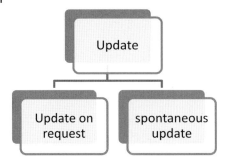

Figure 11.1 Reasons for doing an update.

For a spontaneous update the registrant has to specify in detail what reason made him do the update. Therefore, at least one reason given in Article 22 (1) [2] must be indicated, but it is also possible to provide more than one reason.

11.2
Requested Updates

11.2.1
Update Requested Because of Missing Information

In accordance with Article 20 (2) [4] ECHA will undertake a completeness check of each registration. During the completeness check it will be examined whether all elements required under Articles 10 and 12 or under 17 or 18, as well as the registration fee have been provided [4]. The completeness check shall not include an assessment of the quality or the adequacy of any data or justification submitted [4]. If a registration is incomplete, ECHA will inform the registrant before expiry of the three-week period (remark: three-month period for registrations of phase-in substances submitted in the course of the two-month period immediately preceding the relevant registration deadline mentioned in Article 23 [5]), as to what further information is required in order for the registration to be complete, while setting a reasonable deadline for this [4]. The registrant shall complete his registration and submit it to the Agency within the deadline set [4]. ECHA will confirm the submission date of the further information to the registrant [4]. Once the registration is complete, a registration number will be assigned [6].

However, if the registrant fails to complete his registration within the deadline set, the dossier will be rejected and the registration fee shall not be reimbursed [4]. If a dossier submitted to ECHA did not pass the technical completeness check and the financial completeness check, the corresponding substance cannot be considered as being registered.

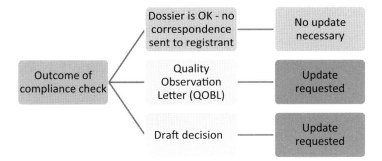

Figure 11.2 Outcome of a compliance and check and necessary further actions.

11.2.2
Updates Requested as a Result of Dossier Evaluation

Dossier evaluation may only be done by ECHA for registrations. Dossier evaluation comprises two different processes: compliance check and examination of testing proposals. Although the processes are different, both may result in ECHA requesting an update of the registration dossier (for details see also Chapter 10 of this book).

11.2.2.1 Update Requested as a Result of a Compliance Check

In a compliance check ECHA can either evaluate the quality of the information in the whole dossier including the chemical safety report or can target the evaluation to a certain part of the dossier [7]. The purpose of the compliance check is to examine whether registration dossiers comply with the requirements of the REACH Regulation [7]. ECHA can decide which dossiers are checked for compliance and whether the examination should cover all or part of a dossier [7]. The REACH regulation requires that the Agency carries out compliance checks on at least 5% of the total number of registration dossiers received for each tonnage band [7, 8].

The outcome of a compliance check (see also Figure 11.2) may be:

- No further action is required.
- A quality observation letter (QOBL) is sent to the registrant, because ECHA identified shortcomings. ECHA informs the registrant in the QOBL about the shortcomings and asks for a revision of the dossier and submission of an updated version [7]. Member States are also informed by ECHA and will take action if the registrant does not clarify the issue [7].
- A draft decision is sent to the registrant if data requirements are not fulfilled. The decision-making process as described by the REACH regulation [1] is followed resulting in a legally binding decision [7].

11.2.2.2 Update Requested after Examination of Testing Proposals

In the examination of testing proposals ECHA evaluates all submitted testing proposals with the aim of checking that adequate and reliable data is produced and to avoid unnecessary vertebrate-animal testing [7]. If a registrant identified data gaps concerning tests or studies foreseen under Annex IX and Annex X [9] for which he submitted testing proposals, ECHA will start a decision-making process. All proposals for tests involving vertebrate animals are published on ECHA's homepage. Everybody who has the requested information can provide it during the period of public consultation [7]. Afterwards, ECHA will evaluate the testing proposals and also information sent by third parties [7]. Based on this ECHA will make a final decision as a result of a decision-making process. The testing proposal may be accepted or rejected. If it seems useful ECHA also may define modified conditions for a test or suggest additional tests to be performed [7]. As time proceeds it is highly recommended to start with all necessary actions immediately after having received a final decision from ECHA or a request for further information, otherwise it may be a problem to meet timelines.

Although doing a certain study may last for example, only 28 days, you may calculate 8 to 12 months for the preparation of the requested update. It may be necessary to look for service providers being in the position to make a test or do a certain study, to ask them for their offers. Evaluation of the offers from different service providers will need some time. After having made the decision to cooperate with a certain provider you may also need the approval of several other persons within your company and therefore extra time has to be calculated for the ordering process. Afterwards, your service provider may have to order animals and further material before the study itself can be started. After having done the study it will often last several weeks until your external partner will be in the position to provide you with the outcome of the test and you will receive a draft report for your review. Further discussions are to be expected. When your company receives the final report you may provide another external service provider, a toxicologist or a consultant company with the study report and ask them to prepare the robust study summary. Again, there will pass some time until the IUCLID Substance data Set is well prepared. Maybe just at the moment you have finalized the dossier, ECHA will release a new version of IUCLID or do maintenance concerning REACH-IT. Therefore, take it seriously, make a detailed plan and try to push the administrative tasks to win some time in case there should occur unexpected difficulties by doing the tests. Maybe a test must be repeated or analytical methods have to be developed.

11.3
Spontaneous Updates

"Following registration, a registrant shall be responsible on his own initiative for updating his registration without undue delay with relevant new information and submitting it to the Agency in the cases" described in Article 22 (1) [2]. Figure 11.3

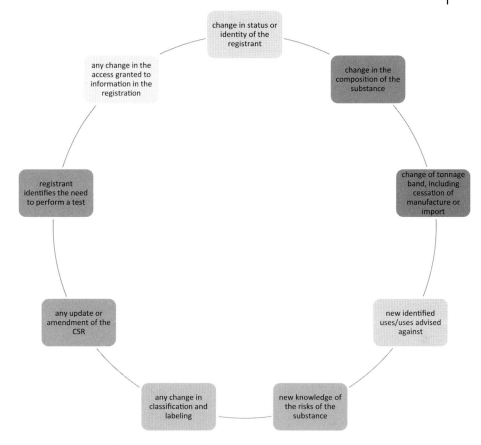

Figure 11.3 Reasons for spontaneous updates.

gives a short overview of the reasons for spontaneous updates as defined in Article 22 (1) (a) to (i) [2].

11.3.1
Update Because of Change in Status or Identity of the Registrant

In Article 22 (1) (a) [10] there is mentioned as a reason for a registrant having to update his dossier "any change in his status, such as being a manufacturer, an importer or a producer of articles, or in his identity, such as his name or address;" A change of the role of the registrant under REACH, for example, former manufacturer now wants also to act as an importer, will result in changing the Substance Data Set in IUCLID, creating a new dossier and afterwards submission of the updated dossier to ECHA.

If there is a change in the identity of the registrant, for example, because the company was sold or the name of the registrant's company changed, then the update must not be done separately at the dossier level for all substances that the registrant registered. It will be enough to indicate this sort of change by updating the information in REACH-IT.

For detailed information concerning changes of legal entity and the resulting consequences see also the REACH-IT Industry User Manual–Part 17 [11].

11.3.2
Update Because of Change in the Composition of the Substance

Article 22 (1) (b) [12] gives as a second reason for a spontaneous update "any change in the composition of the substance as given in Section 2 of Annex VI;" [13].

This means in any case where the registrant manufactures or imports a substance with a composition that is different from the composition given in the registration dossier, ECHA expects the registrant to submit an updated dossier. Smaller deviations may occur when the manufacturing process is changed, a purification step is omitted or a further purification step is added in the manufacturing procedure. Deviations in the composition may also occur if the manufacturer uses a raw material of a different quality, for example, as a consequence of purchasing the raw material from another source. If a raw material is purchased from different sources in different qualities the registrant may check whether the impurity profile of the registered substance could be slightly amended by giving a broader range for the "typical concentration range". It is recommended to give as a lower limit for any impurity "0%" and as the upper limit for the main constituent "100%", although the purity of the registered substance usually is not 100% and the typical concentration of an impurity will be above 0%. By doing so the registrant will avoid having to do an update if the quality of the product increases. If a customer to be demands a product of higher quality and the registrant therefore does a further purification step, this will be covered with the already existing registration. The registrant avoids the necessity of doing an update.

If the impurity profile is different depending on the source of the raw material and/or if an impurity is relevant for classification and labeling because of its concentration in the registered substance, a registrant may also consider to add another separate block for each quality of the substance in IUCLID Section 1.2. Although it seems to be easy to cover different qualities of a product in a registration dossier by doing so, the registrant has to face the fact that several blocks in IUCLID Section 1.2 may have further consequences.

In any case, where also IUCLID Section 1.4 (Analytical information) will be checked by ECHA, the registrant has to provide analytical data concerning each quality or composition given in Section 1.2 of the dossier. This may lead to further costs, especially in cases where ECHA requests further studies/tests at a later stage, because the substance sameness seems no longer to be given by comparison with the quality of the substance used for studies and tests included in the registration dossier.

A further consequence may be the necessity to also add a further block in the Classification and Labeling part (Chapter 2) of the dossier, when the impurity profile of a further Composition leads to a different classification of the substance.

Therefore, each registrant is recommended to consider carefully which figures should be entered in the IUCLID Substance Data Set in Section 1.2 for the Composition to avoid unnecessary problems for the future. Although ECHA does not appreciate registrants providing a broader "typical concentration range" than usually observed, it may be the only possibility to save a lot of time if there are smaller deviations that may occur by accident, for example, by using only once a raw material of a lower quality or from another source or having only once an interruption in the manufacturing process that leads to a slightly different impurity profile.

However, if there occurs the situation that the composition of an already registered substance was not covered in the past, the registrant must not stop the manufacturing process nor the delivery to his customer(s), but shall do the update without undue delay.

11.3.3
Update Because of Change of Tonnage Band

A registrant is also obliged to submit an update in accordance with article 22 (1) (c) [14] in the case of "changes in the annual or total quantities manufactured or imported by him or in the quantities of substances present in articles produced or imported by him if these result in a change of tonnage band, including cessation of manufacture or import;" [14].

If the registrant manufactured and/or imported the requested substance in the last three consecutive calendar years, he may benefit from the fact that the relevant yearly tonnage may be calculated based on the average volume of the three preceding calendar years in accordance with Article 3 (30) [15].

Therefore, sometimes no update will be required although the volume in a single calendar year may be higher than the registered tonnage band.

To ensure that a necessary update can be done without undue delay, it is recommended to do a Substance-Volume Tracking on a regular basis at least once per calendar year for all relevant substances, although "without undue delay" is not defined exactly by the REACH regulation.

Be aware of the fact that for substances that were not manufactured for at least three consecutive years the actual volume has to be considered.

11.3.4
Update Because of New Identified Uses

In Article 22 (1) (d) [16] it is defined that a registrant has the obligation to prepare an update also when there are "new identified uses and uses advised against as in Section 3.7 of Annex VI [17] for which the substance is manufactured or imported;" [16].

For a standard registration in volumes above 10 t/a the registrant has to update not only the corresponding IUCLID Sections 3.5 and 3.6 but also the Chemical Safety Report to be attached in Chapter 13 of the Substance Data Set.

It may be important to add uses advised against especially in cases where a customer intends to use a substance in a way that is not covered by the registration of his supplier, but the supplier is not willing to cover this particular use within his registration. There may be good reasons for such a decision, as it may be time consuming to expand a Chemical Safety Report for a certain use of a single customer and maybe this customer does not want to disclose all necessary details of his manufacturing process to his supplier(s). Another reason may be that a registrant submitted a registration for a transported isolated intermediate in a higher tonnage band, but one or several of his customers intend to use this substance not in accordance with Article 18 (4) [18]. Whenever a registrant sends an update by indicating uses advised against, afterwards the customers may submit their own Downstream User Reports to ECHA concerning such uses.

11.3.5
Update Because of New Knowledge of the Risks of the Substance

The fifth out of nine reasons given for spontaneous updates is mentioned in Article 22 (1) (e) [19]. If a registrant received "new knowledge of the risks of the substance to human health and/or the environment ... which leads to changes in the safety data sheet or the chemical safety report;" [19].

For a standard registration at least the Lead Company has to submit an updated dossier by considering all relevant changes in the IUCLID Chapters 11 and 13. If Members of Joint submission submitted the CSR or Guidance on safe use on their own, they shall not only be informed by the Lead Registrant accordingly, but furthermore will have the obligation to do in addition an update of their own registration dossier. Members of Joint submission that submitted a registration dossier for a transported isolated intermediate will have the obligation to submit an update in each case where the Guidance on Safe Use (IUCLID Chapter 11) is concerned.

As a consequence of the new knowledge of the risks of a substance, there may also be a change in the field of classification and labeling be required. Then it may be necessary to give more than one reason for the update in the header of the registration dossier.

11.3.6
Update Because of Any Change in Classification and Labeling

If there is "any change in the classification and labeling of the substance;" [20], at least the Lead Company has to submit an update of the registration dossier by doing all necessary amendments in IUCLID Chapter 2. If so, Members of Joint submission shall be informed by the Lead Registrant. Members of Joint submission that did an Opt-out concerning the classification and labeling of a substance may also have the obligation to do an update on their own. There can even occur the situation that only a Member of Joint submission has to do such an update for example, if the composition of the substance registered by this Member of Joint submission is different from the composition indicated in the registration dossier of the Lead company and the change in classification and labeling has to be done because of an impurity contained in a concentration that is relevant for the classification.

11.3.7
Update Because of an Amendment in the Chemical Safety Report

There can occur situations that make it necessary to do an "update or amendment of the chemical safety report or Section 5 of Annex VI;" [21]. Section 5 of Annex VI [22] concerns the Guidance on safe use.

Normally such an update is done by the Lead Registrant and afterwards all Members of the Joint submission are provided with all necessary information. If Chemical Safety Report and Guidance on safe use were submitted separately with their own registration dossiers in the past, it will be necessary that they also submit their own updates to ECHA.

11.3.8
Update Because of the Need to Perform Further Tests

A spontaneous update may be submitted to ECHA when "the registrant identifies the need to perform a test listed in Annex IX or Annex X, in such cases a testing proposal shall be developed;" [23]. In most cases such an update will be submitted by a Lead company also on behalf of the Members of a Joint submission but it may also be submitted in the case of single submissions. This sort of update may be necessarily combined with an update of a standard registration because of a tonnage band increase to volumes above 100 t/a.

11.3.9
Update Because of a Change in the Access Granted to Information in the Registration

The last reason for a spontaneous update stated in Article 22 [24] concerns "any change in the access granted to information in the registration" [24].

It can be expected that this type of update will seldom occur and will be of concern for studies and tests or robust study summaries used to prepare the IUCLID Chapters 4 to 7.

Such an update may be required when there is a change of the legal entity of a Data Holder or the registrant no longer has the right to refer to a study because of any reason.

11.4
Update of Dossiers of Formerly Notified Substances

Although the rules for updating a dossier for a former notified substance are the same as for all other registered substances, there can be applied some special rules that are described in Annex 4 of the Data Submission Manual, Part 05 [25]. In general, the former notified substances can be regarded as registered under REACH [26] and the former notifier will have already claimed the registration number by upgrading the former SNIF files into IUCLID5 Substance Data Sets. However, the registrant can benefit from several special arrangements that are granted in accordance with the former notification.

When updating a registration that was previously a notification under Directive 67/548/EEC (NONS) the following two scenarios have to be taken into account [25]:

- tonnage band updates;
- other updates.

11.4.1
Update Because of Tonnage Band Increase for Former Notified Substances

In accordance with Article 24 (2) [27] the registration dossier for a previously notified substance must be updated as soon as the quantity reaches the next tonnage threshold. As substances being subject of notification had to be notified from 10 kg/a and above there is for substances produced in quantities below 1 t/a an update required when reaching the 1 t/a threshold [25].

It is important to be aware of the fact that by doing any update because of increase in tonnage band not only information that is required under REACH [1] for the higher tonnage band, but also any information that corresponds to lower tonnage thresholds has to be provided [25].

For a standard registration at or above 10 t/a a Chemical Safety Report has to be included in Chapter 13 of the registration dossier, unless not required due to reasons given in Article 14 (2) [28]. Furthermore, a CSR must not be submitted when the update does not involve a change in tonnage band and there are no new identified uses [25].

If you are not obliged to submit a CSR, you have to choose in Chapter 13 of the IUCLID Substance Data Set "REACH Chemical Safety report (CSR)" and give the justification for not providing the CSR in the field "Remarks". There are two relevant justifications that could be entered:

- "A CSR is not submitted because it is a previously notified substance which did not reach the next tonnage threshold and which does not fall within the scope of Articles 22 (1) (d) [16], 22 (1) (e) [19] and 22 (1) (f) [20] of the REACH Regulation" [25].

- "A CSR is not submitted because the substance fulfils the requirements of Article 14 (2) [28] of the REACH regulation" [25].

Furthermore, in the dossier header shall be stated in the field "Dossier submission remark" the following derogation statement:

> "This dossier is a registration update of a previously notified substance which did not reach the next tonnage threshold under the REACH regulation. It contains new and updated information." [25]

11.4.2
Other Updates for Former Notified Substances

As we already saw in Section 11.3 of this book, in Article 22 of the REACH regulation [29] are listed in cases when a registrant has to submit an update of his dossier. This also includes updates to include the classification and labeling according to the CLP regulation [30]. In these cases and also in the case of requested updates because of an Agency decision, certain information in the dossier is not required in the case of former notified substances. However, as a minimum the following information should be included:

The new information submitted has to meet all of the TCC requirements, with no special exceptions for previously notified substances [25].

The IUCLID Sections 1.1 and 1.2 have to be complete with the only exception that structural formula that have been submitted on paper under Directive 67/548/EEC are optional [25]. As it does not cause high expenditure of time and may be useful for providing a clearly arranged dossier it is recommended nevertheless to add the structural formula for the reference substances. In IUCLID Section 1.3 "Identifiers" shall be given the notification number (NCD number) as well as the registration number (that was received after claiming the registration number). In Section 1.7 Only Representatives should attach an Appointment letter as usual.

Chapter 2 of the Substance Data Set comprises of Section 2.1 "GHS" that is compulsory since 1 December 2010. In the new version IUCLID5.4 there was introduced Section 2.3 "PBT assessment". If a CSR has to be provided in the dossier, then at least an endpoint summary should be provided in this section of the dossier. If "PBT assessment does not apply" is selected a justification should be given for that [25]. When the dossier is done for a Manufacturer (indicated in Section 1.1), then Section 3.3 "Sites" has to be entered and at least one manufacture site must be linked with a manufacture use given in Section 3.5 [25].

11.4.3
Confidentiality Claims that were Previously Done in the Notification

ECHA normally charges a fee for every single confidentiality request, but there are exceptions for confidentiality requests successfully made under Directive 67/548/EEC, provided that this is confirmed by the registrant in their dossier [31]. For every single Confidentiality flag that already had been in the dossier for the notification under Directive 67/548/EEC and for which the claims were accepted, the registrant is required to write in the justification field adjacent to each confidentiality flag the text "Claim previously made under Directive 67/548/EEC".

This will allow ECHA to invoice correctly and validate the claims already presented under Directive 67/548/EEC [32]. If a registrant forgets to write this justification in each of the relevant fields, that may lead to further fees invoiced by ECHA for each confidentiality flag. Therefore, carelessness will be punished mercilessly concerning the fees.

11.5
Update of Dossiers for PPORD Notifications

Although a PPORD notification is not a registration and therefore has an expiry date, there may be good reasons for doing an update.

First, the situation after the expiry date has to be considered in due time. If the manufacturing process in the meantime has been developed successfully, it may be necessary to do a registration to achieve REACH compliance for the time after the expiry date. For non-phase-in substances this means first the submission of an inquiry dossier and afterwards preparation of an appropriate registration dossier. The procedure corresponds to the usual registration process for non-phase-in substances. The same has to be applied in principle for phase-in substances that were not pre-registered and cannot benefit from a late pre-registration. If the conditions for the PPORD notification are no longer fulfilled before the expiry date the former notifier/registrant to be should start with the necessary tasks earlier. In this case, the former notifier does not do any update of the PPORD notification, but a new registration.

If an extension of the PPORD notification can be justified after the expiry date, because the conditions for PPORD are still met and, furthermore, the notifier did not claim any extension for the substance of request so far, an update of the PPORD notification can be done. The notifying party then may benefit from the exemption for another 5 or 10 years depending on the type of use of the substance, as described in Section 3.9.10 of this book.

For such an update at least in Section 1.3 of the IUCLID Substance Data Set the PPORD notification number shall be added as an identifier.

However, for a PPORD notification there may be further reasons that are similar but not identical to the obligation for doing an update of a registration dossier. The notifier of a substance manufactured or imported for the purpose of PPORD is responsible on his own initiative for updating his notification whenever a change has occurred in the information submitted in accordance with Article 9 (2) [33] and also listed in the supporting Guidance document [34].

Any change in the identity of a PPORD notifier, the tonnage band specified in the PPORD notification, the classification and labeling or in the list of customers is considered relevant [34]. In addition, if the substance composition varies and if

Change in information according to Article 9 (2) (a) to (e) [33]	Required information is specified in the following Section of Annex VI [35]	Relevant Chapter in the IUCLID Substance Data Set
Identity of the manufacturer or importer or producer of articles	Section 1 of Annex VI	1.1 Identification (legal identity section)
Identity of the substance	Section 2 of Annex VI	1.1 General information 1.2 Composition 1.4 Analytical information
Classification of the substance, if any	Section 4 of Annex VI	2.1 Classification and Labeling
Estimated quantity	Section 3.1 of Annex VI	1.9 Product and process-oriented research and development
List of customers, including their names and addresses	–	1.8 Recipients

Figure 11.4 Reasons for updating a PPORD notification.

this is not already described in the notification, then an update must be submitted [34]. An update of the PPORD notification shall have no consequence on the period over which the exemption from registration is valid [34].

An overview of reasons for having the obligation to prepare an update for a PPORD notification is listed in Figure 11.4.

11.6
Costs Concerning Updates

REACH Article 22 (5) demands: "An update shall be accompanied by the relevant part of the fee required in accordance with Title IX" [36]. There, it is determined that "the fees that are required ... shall be specified in a Commission Regulation adopted in accordance with the procedure referred to in Article 133 (3) by 1 June 2008" [37]. Hence, a new piece of law was developed to cover this demand known as "Commission Regulation (EC) No 340/2008 of 16 April 2008 on the fees and charges" that in the meantime was already amended for the first time. Now the "COMMISSION IMPLEMENTING REGULATION (EU) No 254/2013 of 20 March 2013 amending Regulation (EC) No 340/2008" [38] is valid.

Fees will be levied by ECHA for certain types of updates, but there are also updates that can be done free of charge. For an update based on tonnage band increase a registrant has to pay for the update not the complete fees levied by ECHA for this tonnage band, as the already paid fees for the previously registration and maybe earlier updates concerning the tonnage band will be taken into account. However, for a decrease in tonnage band or in case of cessation of manufacturing or import the registrant will receive no refunding by ECHA. Therefore, especially when a company acts in the field of Custom Manufacturing it can be recommended to do the initial registration in a lower tonnage band appropriate to the amounts really manufactured in the beginning and afterwards only if it will be absolutely necessary to do an update for a higher tonnage band. Doing the registration directly in a higher tonnage band may appear attractive in order to save the time for an update, but daily business shows that doing an update at a later stage may not cause as much trouble as it seems. It is important to keep in mind that manufacturing must not be interrupted at the moment the next tonnage threshold is reached and it is also allowed to provide customers with that substance. The only obligation a registrant has is to prepare an update without undue delay. As the REACH regulation does not define a sharp deadline for the submission a registrant may take as much time as seems necessary to prepare the dossier with due diligence. If there are further studies/tests required this may last several months up to years. These conditions will be more acceptable than doing the whole work before starting the manufacturing process for the first time.

For spontaneous updates concerning former NONS, a fee for registration of the substance will be levied if the new tonnage band is higher than the previously notified tonnage band for a standard registration. For intermediates only fees have to be paid if no fee for an on-site isolated intermediate was paid previously. Updating the dossier for a former NONS that can be considered as registered with a full registration at least in the smallest tonnage band of 1 to 10 t/a and for which a registration number already has been claimed will lead to no further fees, as updating of a standard registration to a transported isolated intermediate registration without changing the tonnage band or without further data requirements is free of charge. As for an update from standard registration to an on-site isolated intermediate in any case fees have to be paid (single submission actually 1714 € for a large company) a registrant may think over whether the update can be done from standard registration to a transported isolated intermediate (including the use as an on-site isolated intermediate) instead of doing an update from a standard registration to the on-site isolated intermediate.

11.7
Examples and Exercises

1. Clever AG registered 22 substances. A further 56 substances were pre-registered, but they are not registered yet. Clever AG intends to move:

Actual address:	New address:
Clever AG	Clever AG
Penny Lane 345	Chocolate Street 248
Brussels	Brussels
Belgium	Belgium

How many dossiers do they have to update? Please give a reason for your answer.

☐ 22   ☐ 56   ☐ 78   ☐ other number

2. A registrant intends to do an update of a registration dossier because of tonnage band increase for a certain substance that had been registered in the tonnage band 100 to 1000 t/a. Assume that it is a standard registration and the registrant belongs to a group of large companies.

   What fees can be expected for the update in case of a single submission? What fees can be expected when the registrant is a Member of Joint submission?

3. After the claim of the registration numbers for two previously notified substances, a registrant had a standard registration for each of these substances in the tonnage band of 1 to 10 t/a.

   As ECHA expects the registrant to submit an update for a tonnage band increase even if there are no further data requirements, the registrant intends to submit updated dossiers to ECHA for both of the mentioned substances. Substance A is manufactured in amounts of 200 t/a and can be regarded as an on-site isolated intermediate, whereas substance B is manufactured in amounts of 150 t/a and is used in accordance with REACH Article 18 (4) as a transported isolated intermediate.

   What fees must be expected when submitting the dossier updates for substance A and substance B?

4. A Lead company intends to do a tonnage downgrade from the tonnage band 100 t/a to 1000 t/a to the lowest tonnage band of 1 t/a to 10 t/a because the expected tonnage band of 100 t/a to 1000 t/a never had been reached and they were not able to manufacture and market more than 10 t/a in the past. Now they intend to avoid future demands concerning tests and studies for this substance. The situation at their Co-registrant is similar. This Member of Joint submission did a registration in the tonnage band 10 t/a to 100 t/a but never manufactured more than 10 t/a. How can this tonnage downgrade be done in each case and is there any chance to have the fees that have been paid in the past for the higher tonnage bands refunded?

5. Will doing an update of a PPORD notification for any other reason than extension of the PPORD notification have an influence on the period over which the exemption from registration is valid?

6. Could there be any reason for a spontaneous update that does not require updating a Substance Data Set in IUCLID and hence does not require submission of an updated dossier to ECHA? Could you explain your opinion?

7. What sort of information has to be mentioned in the dossier header of each updated dossier?

References

1. Regulation (EC) No 1907/2006 of the European Parliament and of the Council of 18 December 2006, concerning the Registration, Evaluation, Authorisation and Restriction of Chemicals (REACH), establishing a European Chemicals Agency, amending Directive 1999/45/EC and repealing Council Regulation (EEC) No 793/93 and Commission Regulation (EC) No 1488/94 as well as Council Directive 76/769/EEC and Commission Directives 91/155/EEC, 93/67/EEC, 93/105/EC and 2000/21/EC.
2. See Article 22 (1) of the REACH regulation [1].
3. Guidance on registration, Version 2.0, May 2012, published by the European Chemicals Agency, Chapter 7.
4. See Article 20 (2) of the REACH regulation [1].
5. See Article 23 of the REACH regulation [1].
6. See Article 20 (3) of the REACH regulation [1].
7. Evaluation under REACH, Progress Report 2011, 27.02.2012, published by the European Chemicals Agency, ISBN-13: 1831-6506, ISSN: 978-92-9217-643-3.
8. See Article 41 (5) of the REACH regulation [1].
9. See Annex IX and Annex X of the REACH regulation [1].
10. See Article 22 (1) (a) of the REACH regulation [1].
11. REACH-IT Industry User Manual – Part 17: Legal Entity Change, Version 2.0, July 2012, published by the European Chemicals Agency.
12. See Article 22 (1) (b) of the REACH regulation [1].
13. See Annex VI, Section 2 of the REACH regulation [1].
14. See Article 22 (1) (c) of the REACH regulation [1].
15. See Article 3 (30) of the REACH regulation [1].
16. See Article 22 (1) (d) of the REACH regulation [1].
17. See Annex VI, Section 3.7 of the REACH regulation [1].
18. See Article 18 (4) of the REACH regulation [1].
19. See Article 22 (1) (e) of the REACH regulation [1].
20. See Article 22 (1) (f) of the REACH regulation [1].
21. See Article 22 (1) (g) of the REACH regulation [1].
22. See Annex VI, Section 5 of the REACH regulation [1].
23. See Article 22 (1) (h) of the REACH regulation [1].
24. See Article 22 (1) (i) of the REACH regulation [1].
25. See Annex 4 in the Data Submission Manual, Part 05 – How to complete a technical dossier for registrations and PPORD notifications, Version 3.0, 07/2012, published by the European Chemicals Agency.
26. See Article 24 (1) of the REACH regulation [1].
27. Article 24 (2) of the REACH regulation [1].
28. See Article 14 (2) of the REACH regulation [1].
29. See Article 22 of the REACH regulation [1].
30. See Article 40 of Regulation (EC) No 1272/2008 on classification, labeling and packaging of substances and mixtures (CLP regulation).
31. See Annex 3 in the Data Submission Manual, Part 05 – How to complete a technical dossier for registrations and PPORD notifications, Version 3.0, 07/2012, published by the European Chemicals Agency.
32. Questions and Answers, For the registrants of previously notified substances (Release 6), 22-09-2010, published by ECHA.
33. See Article 9 (2) of the REACH regulation [1].
34. Guidance on Scientific Research and Development (SR&D) and Product and Process Oriented Research and Development (PPORD), (Guidance for the implementation of REACH), February 2008, published by the European Chemicals Agency.

35 See Annex VI of the REACH regulation [1].
36 See Article 22 (5) of the REACH regulation [1] and there cited Title IX of the REACH regulation.
37 See Article 74 (1) of the REACH regulation [1].
38 COMMISSION IMPLEMENTING REGULATION (EU) No 254/2013 of 20 March 2013 amending Regulation (EC) No 340/2008 on the fees and charges payable to the European Chemicals Agency pursuant to Regulation (EC) No 1907/2006 of the European Parliament and of the Council on the Registration, Evaluation, Authorisation and Restriction of Chemicals (REACH).

12
Substances of Very High Concern and Authorization Process

Authorization is a process that is new under REACH [1], whereas Restrictions were already known under former directives. The relevant chapters of the REACH regulation concerning the authorization process are listed as an overview in Figure 12.1.

The aim of authorization is "to ensure the good functioning of the internal market while assuring that the risks from substances of very high concern are properly controlled and that these substances are progressively replaced by suitable alternative substances or technologies where these are economically and technically viable" [2].

Only substances fulfilling criteria of SVHC can be subject of authorization.

12.1
Uses that are Exempted from Authorization

Applications for authorization will in principle have to be done for each use of a certain substance regardless of the quantity of the substance that is manufactured or used. However, there are several uses of a substance that may be exempted from authorization although a particular substance may already be subject of authorization.

Exempted are uses of a substance in a medical device regulated by Directives 90/385/EEC [3], 93/42/EEC [4] or 98/79/EC [5], [6]. In accordance with Article 56 (3) [7] the use of substances in scientific research and development is exempted. Annex XIV [8] shall specify whether an authorization will be required for product and process orientated research and development as well as the maximum quantity exempted [7]. Further exemptions in accordance with Article 56 (4) [9] are uses of substances

- in plant protection products within the scope of Directive 91/414/EEC [10];
- in biocidal products within the scope of Directive 98/8/EC [11];
- as motor fuels covered by Directive 98/70/EC [12] of the European Parliament and of the Council of 13 October 1998 relating to the quality of petrol and diesel fuels;

REACH Compliance – The Great Challenge for Globally Acting Enterprises, First Edition.
Susanne Kamptmann.
© 2014 Wiley-VCH Verlag GmbH & Co. KGaA. Published 2014 by Wiley-VCH Verlag GmbH & Co. KGaA.

> **Chapter 1 Authorization requirement**
> Article 55 Aim of authorization and considerations for substitution
> Article 56 General provisions
> Article 57 Substances to be included in Annex XIV
> Article 58 Inclusion of substances in Annex XIV
> Article 59 Identification of substances referred to in Article 57
>
> **Chapter 2 Granting of authorizations**
> Article 60 Granting of authorizations
> Article 61 Review of authorizations
> Article 62 Applications for authorizations
> Article 63 Subsequent applications for authorization
> Article 64 Procedure for authorization decisions
>
> **Chapter 3 Authorizations in the supply chain**
> Article 65 Obligation of holders of authorizations
> Article 66 Downstream Users

Figure 12.1 Title VII of the REACH regulation deals with Authorization.

- as fuel in mobile or fixed combustion plants of mineral oil products and use as fuels in closed systems.

There may also be exemptions for uses in cosmetic products within the scope of Directive 76/768/EEC [13] and in food contact materials within the scope of Regulation (EC) No 1935/2004 [14], [15].

If a substance that is subject of an authorization is present in a preparation, an authorization application will not be necessary until the limits specified in Article 56 (6) [16] are reached. The limit is 0.1% w/w [17] for substances listed in Article 57 (d), (e) and (f) [18], which means PBT, vPvB and ELOC. For all other substances that were subject of an authorization the lowest concentration limits specified in Directive 1999/45/EC [19] or in Annex I to Directive 67/548/EEC [20] are relevant [21].

Furthermore, in accordance with Article 2 (8) (b) [22] substances that are registered as on-site isolated intermediates or transported isolated intermediates in accordance with Article 18 (4) [23] can benefit from not being subject to an authorization process.

For a short overview on uses that are or may be exempted from authorization see also Figure 12.2. For every use that is or may be exempted you will find also a short hint on the relevant piece of legislation.

12.2
Substances of Very High Concern (SVHC)

Substances of very high concern (SVHC) includes all substances covered by Article 57 [24]. For an overview see Figure 12.3.

12.2 Substances of Very High Concern (SVHC)

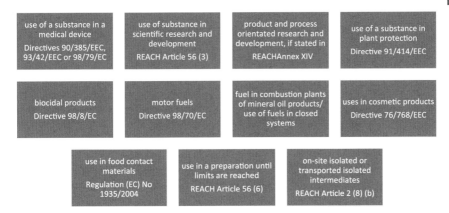

Figure 12.2 Uses of a substance that are or may be exempted from authorization.

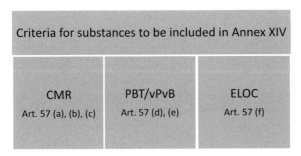

Figure 12.3 Criteria for substances that may be included in Annex XIV of the REACH regulation.

In Article 57 [24] there are listed criteria for substances that may be included in Annex XIV [8] of the REACH regulation. Relevant substances can be

- CMRs Cat. 1 or 2 in accordance with Directive 67/548/EEC [20];
  - carcinogenic substances Cat. 1A or 1B in accordance with Annex I, Section 3.6 of Regulation 1272/2008/EEC [25];
  - mutagenic substances Cat. 1A or 1B in accordance with Annex I, Section 3.6 of Regulation 1272/2008/EEC [25];
  - toxic to reproduction Cat. 1A or 1B in accordance with Annex I, Section 3.6 of Regulation 1272/2008/EEC [25];
- PBT and vPvB substances;
  - substances that are persistent, bioaccumulative and toxic;
  - substances that are very persistent and very bioaccumulative in accordance with the criteria set out in Annex XIII [26];

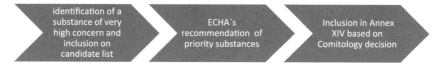

Figure 12.4 Procedure for inclusion of substances in Annex XIV.

- substances – such as those having endocrine disrupting properties or those having PBT or vPvB properties (may be not in accordance with the criteria of Annex XIII, but for which there is scientific evidence of probable serious effects to human health or the environment that give rise to an equivalent level of concern (ELOC) [27].

The procedure to include substances into the authorization system consists of [28]:

- the identification of substances of very high concern and their inclusion on the candidate list with an indication whether they are included in the Agency's work program [28];
- the Agency's preparation of its recommendation of priority substances for inclusion in Annex XIV [8] according to Article 58 (3) [29]; [28]
- a comitology decision on which substances to include in Annex XIV [8]; [28].

This simplified process description is also shown in Figure 12.4 and we will have some more details concerning the relevant steps in the next sections of this book.

12.3
Substance Identification and Identification Procedure

The first step in the authorization process is the identification of substances of very high concern. The procedure to identify SVHC and afterwards inclusion in the candidate list is laid down in Article 59 [30]. To be included in the candidate list of substances of very high concern a substance must have properties as listed in Article 57 [24] and its inclusion in the candidate list is either recommended by a Member State or by ECHA on behalf of the commission.

The identification procedure until eventual inclusion of a substance into Annex XIV [8] shall be described in the next sections as a series of short summaries. One short summary for each single step of the authorization process will form a short overview on this process. This may help to understand the authorization process as a whole. If in depth information is desired the reader may look at the literature to gain more experience.

12.3.1
Identification Procedure

The first step is the preparation of an Annex XV dossier by the authorities. A Member State or the Agency on request by the Commission prepares a dossier in which it provides the argumentation as to why a substance has properties of very high concern [28]. If a substance was included in Annex I of Directive 67/548/EEC [31] it may be enough to reference that.

The dossier shall have the format as described in Annex XV [32]. After having received such a dossier by a Member State ECHA will make the dossier available to all other Member States within 30 days. The non-confidential parts of the dossier will be published on ECHA's website. Every interested party may send in comments on this. The same procedure can be applied if ECHA itself prepared the Annex XV dossier. From the date of circulation Member States and also ECHA may send comments within 60 days [33].

If there are no comments, in the next step ECHA will include the substance of concern in the candidate list. The substance then may be included in Annex XIV at a later stage, if ECHA includes such a substance in its "recommendation on priority substances to the Commission for inclusion into Annex XIV" [28].

If there are any comments either from the authorities (ECHA, Member States) or from any third parties, ECHA will provide the Member State Committee with the dossier within 15 days after expiry of the commenting period. Within a period of a further 30 days the Member State Committee will have to check the situation. If there is a unanimous agreement on the identification, ECHA will include the substance on the candidate list, and may also include it in its recommendations on priority substances to the Commission for inclusion into Annex XIV [28]. If there is no unanimous agreement within the Member State Committee, they will draft their opinion on the issue and sent it to the European Commission. The European Commission then will have to prepare a draft proposal on the identification of the substance within a 3-month period. The final decision will be made in accordance with the comitology procedure [28]. All substances identified in this procedure will form the candidate list of substances from which ECHA may select priority substances for inclusion into the authorization system.

12.3.2
Content of an Annex XV Dossier

An Annex XV dossier is prepared with the objective of the identification of a substance as a CMR (cat 1 or 2) substance, a PBT substance, a vPvB substance or as a substance with probable serious effects that give rise to an equivalent level of concern (ELOC) in accordance with Article 57 [24]. Based on the Annex XV dossier a substance may be proposed to be included in the candidate list for eventual inclusion in Annex XIV [8]. The tasks to be done to prepare an Annex XV dossier

Figure 12.5 Preparation of an Annex XV dossier.

comprise three steps: collecting information, reviewing the information and finally creating a report (Figure 12.5).

The REACH regulation [1] in Annex XV lays down general principles for preparing dossiers to propose and justify [32]:

- harmonized classification and labeling of CMRs, respiratory sensitizers and other effects [34];
- the identification of PBTs, vPvBs, or a ELOC [34];
- restrictions of the manufacture, placing on the market or use of a substance within the Community [34].

In Annex XV (II) [35] there is a short description of the content of these three types of dossiers including hints on what sort of information shall be included concerning "proposal" and "justification" [36]. Concerning restrictions there shall be also analyzed the socio-economic impacts [37] and information on any consultation of stakeholders shall be included in the dossier [37].

The Annex XV dossier consists of two parts that are in any case different from and independent of any registration dossier. A registration dossier consists of a technical dossier in which may be included a Chemical Safety Report (CSR), whereas an Annex XV dossier comprises an Annex XV report and also a technical dossier supporting the Annex XV report [33] (see Figure 12.6). The basic format of the Annex XV report may be adapted to the specific requirements of the individual Annex XV dossier, but it will in any case include a proposal and a justification.

Figure 12.6 Annex XV dossier.

In the following section we will discuss only one type of Annex XV dossier and therefore look at only one type of Annex XV report. We will consider the Annex XV report for the identification of a substance as a CMR, PBT, vPvB or ELOC as this sort of Annex XV dossier may have the greatest consequences concerning the authorization process.

12.3.2.1 Annex XV Report for the Identification of a Substance as a CMR, PBT, vPvB or ELOC

The first part of the Annex XV report contains a summary of the proposal for identification of the substance as a CMR, PBT, vPvB or ELOC [28]. It should also contain the main identifier of the identity of the requested substance and a short summary of the properties of very high concern by considering the relevant properties that are listed in Article 57 [24]. If the substance is already registered, the registration number will be included.

The second part of the Annex XV report contains a detailed technical and scientific justification for the proposal [28]. The format of this section of the Annex XV report follows Part B of the format of the CSR [28]. The format for the technical and scientific justification of the Annex XV report is broken down in a number of sections [28].

In a third part of the Annex XV report information on use, exposure, alternatives and risks may be included [28].

12.4 Inclusion of a Substance in the Candidate List of Substances of Very High Concern (SVHC)

Either ECHA or a Member state recommends the inclusion of a substance in the Candidate list of substances of very high concern. Member states can recommend in principle at any time. First, the Member state informs ECHA about being interested in preparing a dossier for a certain substance. Because of an agreement between Member States and ECHA the dossier submission can be done twice per year (dates in the beginning of a calendar year and another date in the middle of a year). ECHA publishes in the so-called "Registry of Intentions (RoI)" the substance and also the date for submission of the dossier as foreseen by the Member state willing to prepare the dossier for the mentioned substance. ECHA or the Member State prepares the so-called Annex XV Dossier in which it is mentioned which criteria concerning SVHC the certain substance fulfills. If a certain substance is already listed in Annex VI Part II of the CLP regulation [25] this may be a sufficient reason. Afterwards, ECHA provides all other Member States with the dossier within 30 days. The dossier is also published on the ECHA website and within 45 days all interested parties can submit their comments. Member States have a deadline of 60 days for sending their comments. If there are no comments at all, ECHA will include the substance immediately in the candidate list of SVHC. If there are any comments or remarks the Member State Committee (MSC) has to make a decision. In the case of unanimous agreement ECHA will include the substance in the candidate list of substances of very high concern. Only when there is no unanimous agreement within MSC will the European Commission be involved. Within three months they will work out a recommendation with the aim to reach a final decision by the committee procedure [38]. As inclusion of a substance into the candidate list of SVHC is solely based on the intrinsic properties of the substance, inclusion in the candidate list can only be prevented by showing that a certain substance does not fulfill criteria in accordance with Article 57 [24].

12.5
Prioritization and Inclusion of Certain SVHCs in Annex XIV

The number of substances on the candidate list will increase and may be expected to be high in the future, therefore it is useful and necessary to prioritize. Therefore, in the second step of the process ECHA and the Member State Committee (MSC) recommend priority substances from the candidate list to be included in Annex XIV [8]. This means the candidate list of SVHC is a list from which substances can be chosen that should become subject of authorization and therefore should be included in Annex XIV [8]. The procedure for inclusion of substances in Annex XIV [8] is described in Article 58 [39] and Article 133(4) [40].

ECHA makes recommendations every two years which additional substances should be included into Annex XIV. Priority is given to substances that according to Article 58 (3) [29] are

- PBT- or vPvB substances;
- have wide dispersive use;
- produced or imported in high volumes.

ECHA publishes this recommendation on their website and within three months comments can be sent. In summer 2012 there was published the 4th draft recommendation for inclusion in the authorization list. Comments that were sent to ECHA within the set deadline can be considered and the recommendation may be amended by ECHA. In any case, ECHA publishes responses to the comments on their website. There will be a committee procedure for the final decision concerning inclusion in Annex XIV [8].

12.6
Information in Annex XIV

In accordance with Article 58 [39] in Annex XIV [8] is included information concerning the identity of a substance and the relevant intrinsic properties based on Article 57 [24] that lead to the inclusion of the mentioned substance in Annex XIV [8].

Furthermore, there is given a date from which the placing on the market and the use of the substance shall be prohibited ("the sunset date") [41].

Please be aware of the fact that authorization is not dependent on the tonnage band. There is no minimum amount that can be manufactured or imported without having to apply for authorization. This is a great difference compared to substances that only need to be registered when manufactured/imported in amounts of 1 t/a and above.

If a company wishes to continue to use a substance or place it on the market for certain uses after the sunset date, an application must be submitted to ECHA at least 18 months before the sunset date, this is known as the "latest application date" and will also be mentioned in Annex XIV [8].

If the situation should be checked again for certain uses this information will also be included in Annex XIV [8].

In Annex XIV [8] can be given also exemptions from authorization requirement if appropriate in a particular case. In general, substances are all exempted from the authorization process that are exempted from REACH (e.g., radioactive materials, non-isolated intermediates, waste) and in addition there may be further exemptions as mentioned in Section 12.1 of this book.

In February 2011 there were included for the first time in total 6 substances [42] into the former empty Annex XIV [8] and in February 2012 there were included another 8 substances [43]. Therefore, at the end of 2012 there were already 14 substances included.

12.7
Restrictions and Information in Annex XVII

In Annex XVII [44] can be found "Restrictions on the manufacture, placing on the market and use of certain dangerous substances, mixtures and articles".

Within a simple table there are listed in column 1 the "designation of the substance, of the group of substances or of the mixture", whereas in column 2 the corresponding "conditions of restriction" are listed [44].

Within the REACH regulation the Restriction process is defined in Articles 68 to 73 [45]. In Article 68 we find the condition for introducing new and amending current restrictions: "When there is an unacceptable risk to human health or the environment, arising from the manufacture, use or placing on the market of substances, which needs to be addressed on a Community-wide basis, Annex XVII shall be amended in accordance with the procedure referred to in Article 133(4) by adopting new restrictions, or the procedure set out in Articles 69 to 73. Any such decision shall take into account the socio-economic impact of the restriction, including the availability of alternatives. The first subparagraph shall not apply to the use of a substance as an on-site isolated intermediate" [46].

A well-known example of restrictions is mercury. Annex XVII says that it "Shall not be placed on the market: (a) in fever thermometers (b) in other measuring devices intended for sale to the general public (such as manometers, barometers, sphygmomanometers, thermometers other than fever thermometers)" [47]. The restriction shall not apply to measuring devices that were in use in the Community before 3 April 2009. Therefore, there may still be a huge number of households using fever thermometers that contain mercury, but at least they cannot be bought any longer within the European Union.

As additionally stated in Annex XVII concerning mercury: "By 3 October 2009 the Commission shall carry out a review of the availability of reliable safer alternatives that are technically and economically feasible for mercury containing sphygmomanometers and other measuring devices in healthcare and in other professional and industrial uses. On the basis of this review or as soon as new information on reliable safer alternatives for sphygmomanometers and other measuring devices containing mercury becomes available the Commission shall, if appropriate, present a legislative proposal to extend the restrictions in paragraph 1 to sphygmomanometers and other measuring devices in healthcare and in other professional and industrial uses, so that mercury in measuring devices is phased out whenever technically and economically feasible" [47]. For mercury the restriction aims to substitute mercury by any appropriate alternative that bears less risk for human health or the environment, but takes into account that there may be certain uses for which there are no alternatives available yet.

Toluene, which is also listed in Annex XVII, "shall not be placed on the market, or used, as a substance or in mixtures in a concentration equal to or greater than 0.1% by weight where the substance or mixture is used in adhesives or spray paints intended for supply to the general public". In this case, the restriction aims to prevent the use of toluene by consumers. The use by professionals or within industry is in principle not restricted as it is clear that professional users and industry will be aware of the risks and therefore will protect workers in an appropriate way and minimize emissions for example, by the use of appropriate technical means.

12.8
Application for Authorization

The REACH regulation sets up a system under which the use of substances with properties of very high concern and their placing on the market can be made subject to an authorization requirement [28]. This authorization requirement ensures that risks from the use of such substances are either adequately controlled or justified on socio-economic grounds, having taken into account the available information on alternative substances or processes [28]. The burden of proof is placed on the applicant to demonstrate that the risk from the use is adequately controlled or that the socio-economic benefits outweigh the risks [28]. Authorization is required for placing on the market and use(s) of a substance listed in Annex XIV after the sunset date [48]. Applications for authorization can be made by the manufacturer(s), importer(s) and/or Downstream user(s) of the substance, covering one or more uses and/or one substance or a group of substances [48]. An application for authorization has to be done for each relevant use (each use that is not exempted) regardless of the quantity of the substance that is used (see also Figure 12.7).

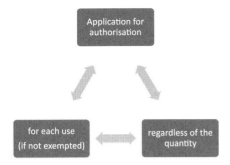

Figure 12.7 Application for an authorization.

12.8.1
Main Elements of an Application for Authorization

In an application for authorization are included basic information concerning identity of the substance and the applicant and also the relevant uses of the requested substance. Furthermore, an analysis of alternatives and if so a substitution plan have to be included. If useful or necessary to receive an authorization at all a socio-economic analysis may be added.

The main elements of an application for authorization are defined in Article 62 (4) [49] and Article 62 (5) [50]. For an overview see Figure 12.8.

Content and volume of the documentation will depend on the availability of appropriate alternatives and also on the fact of whether the application will be compiled based on the Adequate control route or based on the Socio-economic route. An applicant is not urged to fix on only one route, but it will be useful to think through the details beforehand.

The information on substance identity should be based on the Annex XIV entry and on Section 2 of Annex VI of REACH [48].

Applicants can be manufacturer(s), importer(s) and/or downstream user(s) of the substance(s). The main elements that need to be included in the application are similar for each of the above roles of the applicant. Applications can be made by either separate legal entities or a group of legal entities [51].

For each legal or natural person submitting an application the following information is necessary:

- name, address, telephone number, fax number and email address;
- contact person;
- financial and legal identifiers; and
- other relevant contact information [48].

An authorization granted to a downstream user also covers the supply of the substance to the downstream user holding the authorization, irrespective of whether or not the manufacturer(s) or importer(s) have made an application for

Element to be included in an application	Relevant Article of the REACH regulation [1]
In any case:	
Identity of the substance(s)	Article 62 (4) (a)
Name and contact details of the person or persons making the application	Article 62 (4) (b)
A request for authorization, specifying for which use(s) the authorization is sought (including use of the substance in preparations and/or incorporation of the substance in articles, where this is relevant)	Article 62 (4) (c)
Unless already submitted as part of the registration, a CSR covering the risks to human health and/or the environment from the use of the substance(s) arising from the intrinsic properties specified in Annex XIV	Article 62 (4) (d)
Analysis of the alternatives considering their risks and the technical and economic feasibility of substitution	Article 62 (4) (e)
Where analysis shows that suitable alternatives are available, taking into account the elements in Article 60 (5), a substitution plan including a timetable for proposed actions by the applicant.	Article 62 (4) (f)
Optional:	
A socio-economic analysis (SEA)	Article 62 (5) (a)
A justification for not considering risks to human health and the environment arising either from	Article 62 (5) (b)
Emissions of a substance from an installation where a permit was granted in accordance with Directive 96/61/EC; or	Article 62 (5) (b)(i)
Discharges of a substance from a point source governed by the requirements for prior regulation referred to in Article 11 (3) (g) of Directive 2000/60/EC and legislation adopted under Article 16 of that Directive.	Article 62 (5) (b) (ii)

Figure 12.8 Main elements of an application.

an authorization for that use [48]. If the potential applicant is a manufacturer or an importer, he shall keep the downstream users of the substance informed of what use(s) will be covered in the application. Downstream users are recommended to provide information on their specific uses back to the applicant [48]. If the application of a Manufacturer or Importer covers besides their own use(s) also any use(s) for which the substance will be placed on the market, the downstream users themselves do not need to submit their own application.

Each applicant can apply for an authorization for his own use(s) of the substance and/or uses for which the applicant intends to place the substance on the market [48]. As it is up to each supplier (Manufacturer or Importer) to decide whether or not the use(s) of his customers (Downstream users) shall be covered in his authorization application, there may occur cases where a potential applicant does not wish to apply for an authorization for a specific use although he supplies for. Reasons for this may be that the costs for the preparation of the application concerning a particular use are high compared to the profit that can be made by selling a certain product. Another reason may be that the applicant is not in the position to demonstrate safe use and, furthermore, there may be suitable alternatives on the market or it is not possible to demonstrate that the socio-economic benefit is so high that risks to human health or the environment seem to be acceptable. In these cases a Downstream User may wish to compile their own application. This wish may also be justified when the actual use or process used by the downstream user is confidential [48]. If the downstream User can demonstrate adequate control at his site, because he did specific risk management measures and has certain operational conditions in place that may also be a good reason for doing his own application [48]. A higher risk may lie in demonstrating that socio-economic benefits outweigh the risks for the specific use relevant for a Downstream User. However, in cases where are no suitable alternatives available the application for authorization may be granted [48]. If a Downstream User wishes to develop an own application it is useful to inform their supplier(s) and, if relevant also their customers.

As an authorization is granted for the use(s) documented in the CSR, it is important that the description relates to the exposure scenario(s) for that use(s) [48].

All authorization applications need to include a CSR or refer to one submitted as a part of a registration dossier. If a substance was registered in the smallest tonnage band up to 10t/a there was no necessity to prepare a CSR, but for the application for authorization it will be required irrespective of the relevant tonnage.

All applications must include an analysis of alternatives. The purpose of this analysis is to determine if there are any suitable alternative substances or alternative techniques [48]. "When assessing whether suitable alternative substances or technologies are available, all relevant aspects shall be taken into account by the Commission, including:

a) whether the transfer to alternatives would result in reduced overall risks to human health and the environment, taking into account the appropriateness and effectiveness of risk management measures;

b) the technical and economic feasibility of alternatives for the applicant" [52].

There should be considered risks from alternatives [48], technical feasibility of substitution and economic feasibility of substitution [48]. In Figure 12.9 you will find a Checklist that may be used to support you in doing the analysis of alternatives.

If the analysis of alternatives shows that there is any suitable alternative by taking into account reduction of overall risks and technical and economic feasibility of alternatives for the applicant, as a consequence there must be included a substitution plan in accordance with Article 62 (4) (f) [54].

In Figure 12.10 you will find a Checklist that may support you in preparing a substitution plan.

As an authorization can be granted either on the basis of "adequate control" or "socio-economic reasons", there are two different routes [48] to be described:

- the adequate control route (Article 60 (2)); or [56]
- the socio-economic (SEA) route (Article 60 (4)) [57].

> **Check-List to support analysis of alternatives** [53]
>
> - Identification of Annex XIV substance function for the uses applied for
>
> - Identification of possible alternative(s) – substances and/or technologies for the uses applied for
>
> - The assessment of risks to human health and the environment of the alternatives and whether the transfer to alternatives would result in reduced overall risks
>
> - The assessment of the technical feasibility of the alternative(s) for substitution
>
> - The assessment of the economic feasibility of the alternative(s) for Substitution
>
> - The assessment of availability of the alternative(s)
>
> - List of actions required, as well as the time-lines, to switch to an alternative substance/technology
>
> - A justification for the conclusion of the analysis of alternatives if it concludes that there are no suitable alternatives available
>   - If the application is for an Annex XIV substance that cannot be adequately controlled, reference to a SEA
>
> - A justification for selection of the alternative (for an application under the adequate control route)
>   - If the application is for a substance for which adequate control can be demonstrated and there is a suitable alternative available, reference to a substitution plan
>
> - Relevant R & D is documented and explained where appropriate
>
> - References to all information sources cited
>
> - Confidential data is clearly indicated as such

Figure 12.9 Check-List to support analysis of alternatives [53].

12.8.1.1 Adequate Control Route

An authorization can be granted because of adequate control [56] or because of socio-economic reasons [57].

The "adequate control route" applies when it can be demonstrated that the risks to human health or the environment from the use of the substance is adequately controlled in accordance with Section 6.4 of Annex I [58] according to Article 60 (2) [56]; [48].

This route cannot be applied for substances meeting the criteria of CMR [59] or ELOC [27] for which it is not possible to determine a threshold [60]. This means in each case where there is no DNEL or PNEC available the "adequate control route" cannot be applied. Furthermore, this route cannot be applied for PBTs and vPvBs in accordance with Article 57 (d) or (e) [61, 62].

Check-List to support preparation of a substitution plan [55]
The substitution plan includes the following information   • A list of actions   • A timetable for implementation of actions   • The method used to communicate information to stakeholders and the supply chain   •References to supporting information or reports (e.g. SEA)
The list of actions includes   • A series of actions proposed by the applicant (though not always for the applicant to undertake) to facilitate or carry out the substitution   • A proposed timetable with a deadline for the completion of each action   • A justification to present the rationale behind each action/timetable proposed by the applicant   • A reviewing progress against proposed actions/timetable (e.g a GANTT chart)
A substitution timetable should be presented within the plan that   • Contains a start-date for implementation of the substitution plan   • Contains an end-date by which substitution is anticipated to be complete   • Contains a timetable with a deadline for each action   • Is realistic given the limitations identified in the substitution plan   • Contains references to suitable justifications for proposed dates   • Highlights the milestones set within the action plan   • Highlights the internal progress review and internal progress reporting
Internal review of overall substitution position for the purpose of the review report, where relevant   • Are there any new/emerging alternatives that were not present before?   • Is the substitution still the best available option?

Figure 12.10 Check-List to support preparation of a substitution plan [55].

However, when there are DNEL and/or PNEC known it will be possible to demonstrate adequate control by determining with appropriate measurements that these limits are not in any case exceeded.

If there are no limits known authorization may only be granted on the basis of the "socio-economic route".

The application must include:

- CSR (if not already submitted as part of the registration) [48];
- an analysis of alternatives; and [48]
    - where the analysis of alternatives shows that suitable alternatives are available, taking into account elements in Article 60 (5) [52], a substitution plan [48].

12.8.1.2 Socio-economic Assessment (SEA) Route

Based on the above statements the "socio-economic route" has to be applied in any case for PBTs and vPvBs. Furthermore, for all CMRs and if so ELOCs without any DNEL or PNEC the socio-economic route is the only way to have an authorization granted.

It must be demonstrated that the risk to human health or the environment from the use of the substance is outweighed by the socio-economic benefits and there are no suitable alternative substances or techniques [48, 57].

The "SEA route" applies where it can be demonstrated that the risk to human health or the environment from the use of the substance is outweighed by the socio-economic benefits and there are no suitable alternative substances or techniques [48, 57]. It applies in circumstances when adequate control has not been demonstrated and/or for substances meeting the criteria of Article 60 (3) [63]:

- CMR Cat. 1 and 2 substances (Art. 57 (a), (b) or (c)) [59], or substances listed in Annex XIV as being of equivalent concern defined under Article 57 (f) [27], and for which it is not possible to determine a threshold;
- substances that are listed in Annex XIV as being of an equivalent level of concern to PBT or vPvB substances (Article 57 (f)) [27].

The application should include:

- a CSR;
- an analysis of alternatives; and
- a SEA (according to Article 62 (5) [50] optional, but without the granting of authorization on socio-economic grounds is very unlikely [48].

12.9
Data Requirements and Documents Needed for an Application for Authorization

In order to prepare an Authorization application a former registrant is recommended to copy the Substance Data Set used for the registration purposes and afterwards amend this copy. It is not recommended to use the original Substance Data Set as some of the chapters can be omitted for the Authorization application and not all of the uses described in the registration must be subject of the authorization.

The mandatory chapters to be entered in the IUCLID Substance Data Set are listed in Figure 12.11.

12.9.1
Substance Identity and Composition Concerning IUCLID Sections 1.1 and 1.2

The relevant information to be entered in Sections 1.1 and 1.2 concerns substance identity and composition. At least one composition must be indicated. Depending on the type of substance (mono-constituent substance, multi-constituent substance or UVCB) the requested information to be entered may vary. For an overview see Figure 12.12.

Mandatory IUCLID sections		To be attached in this section
1.1	Identification	
1.2	Composition	
1.3	Identifiers	
3.5	Lifecycle Description	Only uses applied for authorization
3.10	Application for authorization of uses	
13	Assessment reports	• CSR • Application form (and other forms)

Figure 12.11 Relevant IUCLID sections in case of an application for authorization.

Mono-constituent substance	Multi-constituent substance	UVCB
• REACH Annex XIV substance entry number		
• Degree of purity of the substance		
• EC number	• EC number (as specified in Annex XIV list)	• EC number of the UVCB substance (as specified in Annex XIV list)
• CAS number	• CAS number	• CAS number
• IUPAC name	• IUPAC name of each constituent	• IUPAC name of each constituent • at least one constituent of the UVCB
• Typical concentration of the constituent	• concentration range of each main constituent	• concentration range of each constituent
If available: • Molecular formula • Molecular weight range • Structural formula	If available for each main constituent: • Molecular formula • Molecular weight range • Structural formula	• Description of the UVCB substance or the process to produce the UVCB (if available)
• IUPAC name of each impurity or additive	• IUPAC name of each impurity or additive	• IUPAC name of each additive
• concentration range of each impurity or additive	• concentration range of each impurity or additive	• concentration range of each additive

Figure 12.12 Information to be entered concerning identity and composition depending on type of substance [64].

12.9.2
Identifiers to be Entered in IUCLID Section 1.3

Section 1.3 is mandatory for the preparation of the following four types of submissions: initial applications, requested updates, subsequent applications and review reports [64].

Identifiers as listed in Figure 12.13 have to be added in the cases given in column 2.

Identifier to be added in Chapter 1.3	When this sort of identifier has to be added
REACH Annex XIV substance entry number	For all types of applications
REACH registration number	ECHA may access a referenced CSR submitted as part of the registration – only when no CSR is submitted with the application for authorization
REACH authorization number	• For subsequent applications • For reviews of authorization

Figure 12.13 Identifiers to be added in IUCLID Section 1.3.

The REACH Annex XIV substance entry number can be found from the Authorization list available at ECHA's website [64].

Please note that if a CSR covering all uses applied for is included in the IUCLID5 dossier when applying for an authorization, no registration number should be provided here [64], whereas in cases when the CSR was submitted with the registration ECHA may access to the referenced CSR when the registration number is provided when applying for an authorization.

The REACH authorization number is relevant for all subsequent applications that refer to an already granted authorization. The same can be applied for reviews of authorization that refer to an already granted authorization [64].

12.9.3
Identification of the "Uses Applied for" Concerning IUCLID Section 3.5

In IUCLID Section 3.5 applicants provide information on the uses they apply for an authorization [64]. The relevant uses can be reported in different tables corresponding to different lifecycle stages. For each relevant lifecycle certain information has to be entered (see Figure 12.14).

For clarity and consistency reasons, the following is important:

- consistent numbering of the uses (IUCLID5 dossier, assessment reports and application web-forms);
- consistent name for each use;
- description of each relevant use (based on the use descriptor system);
- Information about the technical function of the substance;
- For each use in IUCLID Section 3.5 that is also listed in Section 3.10, a reference to (a) relevant exposure scenario(s) shall be made.

Life cycle stage	Fields to be filled in in IUCLID Section 3.5 (minimum information)
Formulation	Identified use name
	Process category
	Technical function
Uses at industrial sites	Identified use name
	Process category
	Technical function (if not available, a description shall be given in the Remarks field)
	Environmental release category
	Subsequent service life relevant for that use? (if not available alternatively: Sector of end use)
Uses by professional workers	Identified use name
	Process category
	Technical function (if not available, a description shall be given in the Remarks field)
	Environmental release category
	Subsequent service life relevant for that use? (if not available alternatively: Sector of end use)
Consumer uses	Identified use name
	Technical function (if not available, a description shall be given in the Remarks field)
Article Service Life	Service life name
	Technical function (if not available, a description shall be given in the Remarks field)

Figure 12.14 Life cycle stages and information to be entered.

12.9.4
Assessment Reports Concerning IUCLID Chapter 13

In IUCLID Chapter 13 different types of assessment reports can be attached [64]. All applications contain or refer to the Chemical Safety Report and contain the application form (generated by a web-form that can be found on ECHA's homepage). Any additional reports that cannot be attached in the other IUCLID chapters should be included in Chapter 13. This may be, besides the application form also the concordance table and if so a mapping of changes performed.

12.9.5
Information to be Provided in the Dossier Header

For subsequent applications that refer to a submitted application and for any update requested by the authorities the last submission number has to be provided in the dossier header [64].

12.10
Submission of the Application of Authorization, Deadlines and Fees

Applications for authorization shall be submitted to ECHA in accordance with Article 62 (1) [65]. The submission has to be done via ECHA's website. Please take note of the fact that it is not possible to submit applications for authorization via REACH-IT.

The deadlines for applications for authorization will be set by the Commission for each substance when it is listed in Annex XIV [48]. Submission of an application for an authorization shall be done 18 months prior to the sunset date at the latest.

ECHA establishes specific windows for submitting applications for authorization. Depending on the corresponding latest application date the submission window foreseen by ECHA and also published on its website will be a two-week period about 3 months previous to the corresponding latest application date.

It can be estimated that it will take 12 months up to 2 years to prepare a new application, therefore applicants to be should start their activities in time.

As Article 62 (7) demands "an application for an authorization shall be accompanied by the fee required in accordance with Title IX" [66]. The fees were specified in Commission Regulation EC 340/2008, which was amended in March 2013. Now the COMMISSION IMPLEMENTING REGULATION (EU) No 254/2013 of 20 March 2013 amending Regulation (EC) No 340/2008 [67] is valid. Already in accordance with Article 8 of Commission Regulation EC 340/2008 [68] and Article 9 of Commission Regulation EC 340/2008 [69] the fees have to be paid for applications for an authorization and there are also charges for reviews of authorizations (Figure 12.15).

Base fee	EUR 53 300
Additional fee per substance	EUR 10 660
Additional fee per use	EUR 10 660
Additional fee per applicant	Additional applicant is not an SME: EUR 39 975
	Additional applicant is a medium enterprise: EUR 29 981
	Additional applicant is a small enterprise: EUR 17 989
	Additional applicant is a microenterprise: EUR 3998

Figure 12.15 Standard fees for applications for an authorization for large companies [70]. Remark: Fees were again reduced for medium, small and microenterprises in the last amendment of the "fees regulation" [67], whereas the fees to be paid by large companies has increased compared to the former valid regulation.
Charges for the review of an authorization have to be paid at the same level as the fees for application for an authorization.

12.11
Subsequent Applicants and their Obligations

In accordance with Article 63 (1) to (2) [71] if an application has already been made for a use of a substance and an authorization has been granted, a subsequent applicant may refer to the appropriate parts of the previous application provided that the subsequent applicant has the permission from the previous applicant to refer to these parts of the application [72]. In any case, the subsequent applicant shall update the information of the original application as necessary [73].

If an authorization has been granted for a use of a substance, a subsequent applicant may refer to the appropriate parts of the previous application submitted in accordance with Article 62 (4) (d), (e), (f) [74] and Article 62 (5) (a) [75], provided that the subsequent applicant has permission from the holder of the authorization to refer to these parts of the application [76].

In both cases [72] and [76] a subsequent applicant can refer to the following parts of the application (provided he has the permission from the previous applicant or authorization holder) [48]:

- CSR
- analysis of alternatives;
- substitution plan;
- socio-economic analysis [48].

In accordance with Article 63(3) [73] the subsequent applicant shall update the information of the original application as necessary [73].

Furthermore, there is requested the following additional information by the subsequent applicant [48]:

- general application information [48];
- substance identity (should relate to the substance used by the subsequent applicant) [48];
- request for authorization for specific use(s) (can refer to the previous applicant's CSR, SEA or analysis of alternatives and substitution plan as appropriate) [48];
- other information (if appropriate) [48].

IUCLID Sections 1.1, 1.2 and 1.3 have to be submitted again. All "uses applied for" in a subsequent application must be listed in Section 3.10 of the IUCLID Substance Data Set. Please do not forget to attach the application form in Chapter 13 [64].

12.12
Process after Submission of the Application for Authorization

Previous to granting an authorization there may occur a situation in which an applicant is urged to do an update and there may also be situations in which he will want to do an update on his own initiative. However, the options are not similar to those for a registration.

12.12.1
Requested Update

As an answer on a request of the Committees in accordance with Article 64 (3) [77] an update has to be submitted to ECHA [64]. The IUCLID5 dossier needs to be updated and in Chapter 13 of the IUCLID dossier a document shall be attached indicating exactly where updates have been made. In the dossier header you have to enter the last submission number and the annotation number (mentioned in the request letter), the rest of the header in principle can be done as for any update of a registration.

12.12.2
Spontaneous Update

It is of high importance to know that spontaneous updates in the case of authorization applications are not allowed at all and therefore will not be processed by ECHA.

12.12.3
To Dos after Granting of an Authorization

The authorization number will be published in the Official Journal, but obligations for the holder of an authorization are not complete at that moment.

Within the deadline specified in the decision a holder of authorization shall update the registration dossier in accordance with Article 22 (2) [78] to take account of the granted authorization.

The safety Data Sheet shall also be updated without delay as determined in Article 31 (9) (b) [79].

In accordance with Article 65 [80] holders of an authorization shall include the authorization number on the label before they place the substance or a preparation containing the substance on the market for an authorized use. This shall be done without delay once the authorization number has been made publicly available [80]. The same has to be done by Downstream Users.

Furthermore, Downstream Users have the obligation to notify the Agency within three months of the first supply of the substance [81].

12.12.4
To Dos after Refusal of an Authorization

If an authorization request is refused, the applicant needs to update the registration taking into account the decision [78] within the deadline specified in the decision [48] and also the Safety Data Sheet shall be updated in accordance with Article 31 (9) (b) [79].

12.12.5
Review of Authorizations

Authorizations granted in accordance with Article 60 [82] shall be regarded as valid until the Commission decides to amend or withdraw the authorization in the context of a review, provided that the holder of the authorization submits a review report at least 18 months before the expiry of the time-limited review period [61].

Authorizations granted for certain uses will be subject to a review period [48, 61]. Authorizations may be reviewed at any time if the circumstances of the original authorization have changed so as to affect the risk to human health or the environment, or the socio-economic impact, or if new information on possible substitutes becomes available [84].

During such a review the Commission may decide to amend or withdraw the authorization [85].

If there is new information on possible substitutes it will apply to both routes of authorization – "Adequate control route" and also to the "Socio-economic route". In order to continue to benefit from an authorization the holder must submit a review report at least 18 months before the expiry of the time-limited review period [48]. The review report shall cover only the parts of the original application that have changed [86].

In the case of a review report the reviewed/updated IUCLID5 dossier has to be submitted again to ECHA. In Section 1.3 the previous authorization number has to be provided. In Chapter 13 a document shall be attached describing those parts that have been updated. By creating the dossier "The submission is a review report" shall be ticked.

12.13
Examples and Exercises

1. Can a substance solely used as a solvent become subject of an authorization?

2. A registrant registered a phase-in substance A as a transported isolated intermediate in 2010. Substance A meets the criteria for classification as carcinogenic category 1 in accordance with Directive 67/548/EEC. When will the registrant be required to prepare an authorization application for substance A?

3. A substance is used as an on-site isolated intermediate by a company A, another company B sells the same substance to consumers. Which company may be concerned if this substance is included in Annex XVII?

4. You intend to prepare a IUCLID Substance Data Set that shall be used for the purpose of application for authorization. As you already did a registration for

that substance including use as a solvent, use as a catalyst, use as a transported isolated intermediate and use as an on-site isolated intermediate you intend to use the former registration dossier to copy the relevant entries from Section 3.5 of the Substance Data Set. Will you be able to use the Substance Data set as it had been used for the registration dossier or will you amend anything?

References

1 Regulation (EC) No 1907/2006 of the European Parliament and of the Council of 18 December 2006, concerning the Registration, Evaluation, Authorisation and Restriction of Chemicals (REACH), establishing a European Chemicals Agency, amending Directive 1999/45/EC and repealing Council Regulation (EEC) No 793/93 and Commission Regulation (EC) No 1488/94 as well as Council Directive 76/769/EEC and Commission Directives 91/155/EEC, 93/67/EEC, 93/105/EC and 2000/21/EC.
2 See Article 55 of the REACH regulation [1].
3 COUNCIL DIRECTIVE of 20 June 1990 on the approximation of the laws of the Member States relating to active implantable medical devices (90/385/EEC).
4 COUNCIL DIRECTIVE 93/42/EEC of 14 June 1993 concerning medical devices.
5 DIRECTIVE 98/79/EC of the European Parliament and of the council of 27 October 1998 on in vitro diagnostic medical devices.
6 See Article 62 (6) of the REACH regulation [1].
7 See Article 56 (3) of the REACH regulation [1].
8 See Annex XIV "List of substances subject to authorization" of the REACH regulation [1].
9 See Article 56 (4) of the REACH regulation [1].
10 COUNCIL DIRECTIVE of 15 July 1991 concerning the placing of plant protection products on the market (91/414/EEC).
11 DIRECTIVE 98/8/EC OF THE EUROPEAN PARLIAMENT AND OF THE COUNCIL of 16 February 1998 concerning the placing of biocidal products on the market.
12 Directive 98/70/EC of the European Parliament and of the Council of 13 October 1998 relating to the quality of petrol and diesel fuels and amending Directive 93/12/EEC.
13 COUNCIL DIRECTIVE of 27 July 1976 on the approximation of the laws of the Member States relating to cosmetic products (76/768/EEC).
14 REGULATION (EC) No 1935/2004 OF THE EUROPEAN PARLIAMENT AND OF THE COUNCIL of 27 October 2004 on materials and articles intended to come into contact with food and repealing Directives 80/590/EEC and 89/109/EEC.
15 See Article 56(5) of the REACH regulation [1].
16 See Article 56 (6) of the REACH regulation [1].
17 See Article 56 (6) (a) of the REACH regulation [1].
18 See Article 57 (d), (e) and (f) of the REACH regulation [1].
19 DIRECTIVE 1999/45/EC OF THE EUROPEAN PARLIAMENT AND OF THE COUNCIL of 31 May 1999 concerning the approximation of the laws, regulations and administrative provisions of the Member States relating to the classification, packaging and labeling of dangerous preparations.
20 COUNCIL DIRECTIVE of 27 June 1967 on the approximation of laws, regulations and administrative provisions relating to the classification, packaging and labeling of dangerous substances (67/548/EEC).
21 See Article 56 (6) (b) of the REACH regulation [1].
22 See Article 2 (8) (b) of the REACH regulation [1].

23 See Article 18 (4) of the REACH regulation [1].
24 See Article 57 of the REACH regulation [1].
25 REGULATION (EC) No 1272/2008 OF THE EUROPEAN PARLIAMENT AND OF THE COUNCIL of 16 December 2008 on classification, labeling and packaging of substances and mixtures, amending and repealing Directives 67/548/EEC and 1999/45/EC, and amending Regulation (EC) No 1907/2006.
26 See Annex XIII of the REACH regulation [1].
27 See Article 57 (f) of the REACH regulation [1].
28 Guidance for the preparation of an Annex XV dossier on the identification of substances of very high concern, June 2007, published by the European Chemicals Agency.
29 See Article 58 (3) of the REACH regulation [1].
30 See Article 59 of the REACH regulation [1].
31 See Annex I of [20].
32 See Annex XV of the REACH regulation [1].
33 Guidance for the preparation of an Annex XV dossier for restrictions, June 2007, published by the European Chemicals Agency.
34 See Annex XV (I) of the REACH regulation [1].
35 See Annex XV (II) of the REACH regulation [1].
36 See Annex XV (II) 1 to 3 of the REACH regulation [1].
37 See Annex XV (II) 3 of the REACH regulation [1].
38 See Article 133 of the REACH regulation [1].
39 See Article 58 of the REACH regulation [1].
40 See Article 133 (4) of the REACH regulation [1].
41 See Article 58 (1) (c) (i) of the REACH regulation [1].
42 COMMISSION REGULATION (EU) No 143/2011 of 17 February 2011 amending Annex XIV to Regulation (EC) No 1907/2006 of the European Parliament and of the Council on the Registration, Evaluation, Authorisation and Restriction of Chemicals ("REACH").
43 COMMISSION REGULATION (EU) No 125/2012 of 14 February 2012 amending Annex XIV to Regulation (EC) No 1907/2006 of the European Parliament and of the Council on the Registration, Evaluation, Authorisation and Restriction of Chemicals ("REACH").
44 See Annex XVII Restrictions on the manufacture, placing on the market and use of certain dangerous substances, mixtures and articles of the REACH regulation [1].
45 See Article 68 to 73 "Restrictions process" of the REACH regulation [1].
46 See Article 68 of the REACH regulation [1].
47 See Annex XVII, 18a. "Mercury" of the REACH regulation [1].
48 Guidance on the preparation of an application for authorization, Version 1, January 2011, published by the European Chemicals Agency.
49 See Article 62 (4) of the REACH regulation [1].
50 See Article 62 (5) of the REACH regulation [1].
51 See Article 62 (2) of the REACH regulation [1].
52 See Article 60 (5) of the REACH regulation [1].
53 See Appendix 3 of [48].
54 See Article 62 (4) (f) of the REACH regulation [1].
55 See Appendix 6 of [48].
56 See Article 60 (2) of the REACH regulation [1].
57 See Article 60 (4) of the REACH regulation [1].
58 See Section 6.4 of Annex I of the REACH regulation [1].
59 See Article 57 (a), (b) and (c) of the REACH regulation [1].
60 See Article 60 (3) (a) of the REACH regulation [1].
61 See Article 57 (d) and (e) of the REACH regulation [1].
62 See Article 60 (3) (b) of the REACH regulation [1].
63 See Article 60 (3) of the REACH regulation [1].
64 Data Submission Manual, Part 22 – How to Prepare and Submit an Application for

Authorisation using IUCLID 5, Version 2.0, July 2012, published by the European Chemicals Agency.
65 See Article 62 (1) of the REACH regulation [1].
66 See Article 62 (7) of the REACH regulation [1].
67 COMMISSION IMPLEMENTING REGULATION (EU) No 254/2013 of 20 March 2013 amending Regulation (EC) No 340/2008 on the fees and charges payable to the European Chemicals Agency pursuant to Regulation (EC) No 1907/2006 of the European Parliament and of the Council on the Registration, Evaluation, Authorisation and Restriction of Chemicals (REACH).
68 See Article 8 of the COMMISSION REGULATION (EC) No 340/2008 of 16 April 2008 on the fees and charges payable to the European Chemicals Agency pursuant to Regulation (EC) No 1907/2006 of the European Parliament and of the Council on the Registration, Evaluation, Authorisation and Restriction of Chemicals (REACH).
69 See Article 9 of the COMMISSION REGULATION (EC) No 340/2008 of 16 April 2008 on the fees and charges payable to the European Chemicals Agency pursuant to Regulation (EC) No 1907/2006 of the European Parliament and of the Council on the Registration, Evaluation, Authorisation and Restriction of Chemicals (REACH).
70 See Annex VI, Table 1 of the COMMISSION IMPLEMENTING REGULATION (EU No 254/2013) [67].
71 See Article 63 (1) to (2) of the REACH regulation [1].
72 See Article 63 (1) of the REACH regulation [1].
73 See Article 63 (3) of the REACH regulation [1].
74 See Article 62 (4) (d), (e), (f) of the REACH regulation [1].
75 See Article 62 (5) (a) of the REACH regulation [1].
76 See Article 63 (2) of the REACH regulation [1].
77 See Article 64 (3) of the REACH regulation [1].
78 See Article 22 (2) of the REACH regulation [1].
79 See Article 31 (9) (b) of the REACH regulation [1].
80 See Article 65 of the REACH regulation [1].
81 See Article 66 (1) of the REACH regulation [1].
82 See Article 60 of the REACH regulation [1].
83 See Article 61 of the REACH regulation [1].
84 See Article 61(2) (a) and (b) of the REACH regulation [1].
85 See Article 61(3) of the REACH regulation [1].
86 See Article 61 (1) of the REACH regulation [1].

13
Achieving REACH Compliance within Your Company – How to Implement Processes to Ensure Legal Compliance

Within the previous chapters of this book we learned the basic principles concerning obligations several actors within a supply chain may have under REACH [1] and what has to be done to achieve REACH compliance concerning several aspects of REACH. Now it seems to be useful to have a sort of summary by having the view on a single enterprise and the processes to be implemented within this company to achieve REACH compliance in this company. Within this chapter I will present some lists and tables that may help to have an overview concerning the particular demands for your own company. Therefore, here the focus will lie on processes to be worked on within your own company, whereas in Chapter 14 we will open the view to the communication up and down in the supply chain.

First, it will be of high interest to find out whether your company has registration obligations for certain substances that are either used or manufactured within your company. As the procedures to be implemented may be different for raw materials and substances manufactured within a particular company there will be two separate sections dedicated to raw materials and substances produced within your company. Where useful and necessary there will also be made a distinction between necessities concerning EU and Non-EU companies.

13.1
List of Used Raw Materials

To find out what sort of obligations your company may have under REACH [1] concerning a particular substance that is supplied to your company as a raw material, it is necessary to list all raw materials used in your company in connection with each supplier for a particular substance. To define further obligations it may be necessary to communicate with the supplier. Figure 13.1 shows the minimum of involved parties to ensure REACH compliance concerning raw materials. There may be further departments that will be involved in your company and also at your supplier's site. Your supplier may be a registrant but it is also possible that there will be a manufacturer up the supply chain acting as the registrant or there is no registrant at all and it will be up to your own company to become the registrant.

REACH Compliance – The Great Challenge for Globally Acting Enterprises, First Edition.
Susanne Kamptmann.
© 2014 Wiley-VCH Verlag GmbH & Co. KGaA. Published 2014 by Wiley-VCH Verlag GmbH & Co. KGaA.

Figure 13.1 Parties involved in the process to ensure REACH Compliance concerning raw materials.

13.1.1
Define Your Role under REACH

To define the role of your company, it is important to know where the supplier is located and whether he takes care of the registration obligations or not.

If you are responsible for the physical introduction of substances or preparations into the EU, you have the role of an importer under REACH and you may have to register the substances [2]. If you purchase from a supplier in another EU country you are not an importer and do not have to register [2]. If you purchase substances or preparations from a Non-EU supplier who has an Only Representative, you are a downstream user under REACH and you do not have to register [2].

In Figure 13.2 you will find a form that may be used to define the role of your company concerning a particular substance. We assume in this case that your company is located within the EU. When a substance is purchased from more than only one supplier the situation needs to be defined for each supplier and the amounts purchased from these supplier.

If your company is located outside of the EU the demands are slightly different, as there are only registration obligations for raw materials that later on shall be delivered to the EU. A table that may support you as a Non-EU company can be found in Figure 13.3. As Substance 1 purchased by the Non-EU company will not be forwarded to the EU they may purchase this substance from every supplier as in the past. Concerning substance 2 that for example, may be a solvent that will be used to precipitate a solution of a substance in this solvent the situation may be different as this substance 2 has to be registered for the supply of EU

Substance name	CAS	EC	Supplier	Supplier located in the EU/Non-EU	Is there an OR for Non-EU supplier	Has supplier (OR) (pre)registered	Amounts of substance purchased by this supplier	Role of our company concerning this substance	Registration obligation/To Dos for our company
Substance 1	...	...	A	EU	-	yes	15 t/a	DU	-
			B	Non-EU	no	no	9 t/a	I	Registration obligation
			C	Non-EU	yes	yes	20 t/a	DU	-
			D	Non-EU	no	no	5 t/a	I	Registration obligation
Substance 1 – Total tonnage to be considered for own (pre-)registration							14 t/a	I	(Pre-) registration in tonnage band 10 to 100 t/a

Figure 13.2 Define the role of your company located within the EU concerning raw materials supplied to you.

TASK: Define your company's role under REACH

1. Make a list including all raw materials (substances, preparations) purchased by your company.
2. Add for each substance all suppliers and define where the supplier is located (within the EU or outside of the EU).
3. EU suppliers have to take care of the registration obligations therefore you will be downstream user (DU) for substances purchased from EU suppliers.
4. For Non-EU suppliers find out whether they take care of registration obligations via an Only Representative or whether they do not. Depending on the outcome you will be either downstream user or an importer for this substance purchased from the corresponding Non-EU supplier.
5. Define whether you are willing to act as an importer under REACH or whether you will rather cooperate only with suppliers that take care of the registration obligations.

Substance name	CAS	EC	Supplier	Supplier located in the EU/Non-EU	Is there an OR for Non-EU supplier	Has supplier (OR) (pre-)registered	Amounts of substance purchased by this supplier	Is substance used outside of the EU or will it be forwarded to EU customer	Is it necessary to buy that raw material from a registered source?	Remarks
Substance 1	...	...	A	EU	–	yes	15 t/a	Non-EU	no	✓
			B	Non-EU	no	no	9 t/a	Non-EU	no	✓
			C	Non-EU	yes	yes	20 t/a	Non-EU	no	✓
			D	Non-EU	no	no	5 t/a	Non-EU	no	✓
Substance 1 can be supplied from all suppliers without having REACH relevance										
Substance 2			A	EU	–	yes	5 t/a	Forward to the EU	yes	✓
			B	Non-EU	no	no	5 t/a			No supplies for the requested purpose
Substance 2 must be registered – therefore buy substance 2 in future only from registered source as supplier A										

Figure 13.3 Define the registration obligations concerning raw materials purchased by a Non-EU company.
TASK: Define as a Non-EU company whether purchased raw materials are appropriate to meet the demands of REACH in case certain substances are forwarded by your company to EU customers

1. Make a list including all raw materials (substances, preparations) purchased by your company.
2. Mark all raw materials that you purchase and afterwards forward as such to EU customers (e.g., solvents used at first in a manufacturing process and that are afterwards sold as a preparation to EU customers).
3. Add for each relevant substance all suppliers and define where the supplier is located (within the EU or outside).
4. For Non-EU suppliers add whether they have an OR that will take care of the registration obligations under REACH and if so, please ensure that they are willing to cover also amounts of the substance that are delivered to your company and afterwards will be forwarded into EU.
5. If you intend to forward a substance purchased as a raw material from any source to EU customers later on, please ensure that you buy the requested substance from a (pre-)registered source. If you are not in the position to purchase from a supplier that will ensure that you are a downstream user for such a substance, please inform your EU customers that they will have to take care of the registration obligations in the role of an importer. (Remark: Your company in this case will act as a Non-EU trader and therefore will not be allowed to register this substance via an Only Representative within the EU even if you would be willing to do so.)

customers. Either the EU customers that order this solution from the Non-EU manufacturer have to take care of the registration obligations or the Non-EU manufacturer buys the solvent from a registered source (e.g., an EU manufacturer of the solvent) and afterwards the customer buying the solution from the Non-EU manufacturer may benefit from the fact that the solvent is already registered and for the (re-)import there are no further registration obligations when it can be ensured that the (re-)imported substance 2 is the same as the substance 2 supplied to the Non-EU manufacturer.

A Non-EU company has to define whether a supplier may be appropriate to make the customers of the Non-EU manufacturer benefit from being Downstream Users in certain situations. An EU company may consider the same if they are not interested in becoming registrants themselves.

13.1.2
Define the Registration Deadline Based on Properties

For pre-registered substances the registration deadline is connected not only to the tonnage band, but also to properties as R50/53 and CMR. If a substance is on the candidate list for SVHC it may have consequences on availability and allowed uses, therefore a column for SVHC is included in the below list in Figure 13.4.

The list aims to support your company when a raw material shall be registered by your company in the role of the importer.

However, the list may be also adapted for supervising further actions to be done by the suppliers. It is important to know the registration deadlines even when your own company does not intend to register, because the registration deadline may cause inconveniences. Availability of a substance may be influenced if there will be a cessation of manufacture after the deadline, because of the high costs to be expected for the preparation of a registration dossier. Furthermore, it may have consequences concerning the handling of a substance as soon as it is registered. Substances registered as transported isolated intermediates have to be used in accordance with Article 18 (4) of the REACH regulation for the synthesis of another substance. For substances registered with a standard registration the supplier will have to provide an eSDS. In each case it may be useful to keep an eye on the tasks to be done by the suppliers and if there is no communication down the supply chain it may be necessary to request certain information from the supplier(s). Please keep in mind that non-phase-in substances must be registered previous to manufacture or import of 1 t/a and above.

13.1.3
Identify Uses of a Certain Substance within Your Company

It is of high importance to clarify how a substance is used within your company. There may be more than one use for a certain substance. You may use the form in Figure 13.5 to list the uses of a raw material within your company.

Substance	CAS	EC	R50/53	CMR	Included in SVHC candidate list *)	Tonnage band			Covering amounts purchased from supplier	Registration deadline
						1–10 t/a	10–100 t/a	100–1000 t/a	>1000 t/a		
Substance 1	…	…	x			x				A, B, C	2018
Substance 2	…	…	x					x		C	2010
Substance 3	…	…		x		x				B	2010
…											
Substance x	…	…	–	–			x			D, E	2018

Figure 13.4 Properties of raw materials and corresponding registration deadline for pre-registered substances.
*) it could also be checked whether a certain substance is subject of restrictions or subject of authorization by checking the relevant Annnexes of the REACH regulation.

- A company located outside of the EU may use the above list slightly amended with the aim to be aware until when their EU-suppliers will have to register a substance or may terminate manufacturing of a certain substance.

TASK: Define registration deadline based on properties of a substance and relevant tonnage band and ensure availability of raw materials needed within your company

1. Make a list including all raw materials purchased by your company
2. Add which substances are CMR substances or have properties that lead to R50/53. These will have a direct influence on the registration deadline for pre-registered substances.
3. You may also add a note if a substance is included in the candidate list for substances of very high concern, in Annex XVII (subject of Restriction) or Annex XIV (subject of Authorisation) of the REACH regulation. This information may help to define whether there could be a high risk that the substance will not be available any more in future or whether costs for purchasing may increase because of an expensive authorization process.
4. In all cases where your company does not intend to register yourself, you are recommended to communicate with your supplier(s) with the aim to find out whether your supplier intends to register and if so when he will do that. You may also ask your supplier(s) whether REACH will have an influence on the availability of a certain raw material purchased from this supplier in future.
5. Define further tasks to be done by your company based on the available information.

Substance	CAS	EC	Type of registration required		Uses described BY Use descriptors (1 line for each use)					To Do
			Standard registration	TII	ERC	PROC	PC	Technical function of the Substance	SU	
Substance 1	...	...	x		...	...	...	...	...	
				x	6a	1	19	Intermediate	8, 9	
			x			4	19	Intermediate, but not under SCC		
			x					Catalyst		
			x					Monomer for manufacturing of a polymer		
			x					solvent		

Figure 13.5 Identify uses of a certain substance within your company.
TASK: Define type of registration that is required based on the use of a certain substance within your company
1. Make a list including all raw materials purchased by your company.
2. Determine how each substance is used within your company and whether there will be a registration as a transported isolated intermediate appropriate or whether a standard registration will be required.
3. Define your uses by using the Use Descriptor system.
4. You may forward your uses to the supplier(s) to make them identified uses and ask your supplier to cover these uses in his registration or to forward them to the next actor up the supply chain with the aim to have them covered in the registrant's registration dossier.
5. Alternatively: If your company will register in the role of the importer you may use this information for the preparation of the registration dossier and if so for the compilation of exposure scenarios and a Chemical Safety Report.

Further tasks and processes where the uses identified within your company are relevant:
- Check SDS/eSDS provided by a supplier to find out whether your own uses are covered with registration of the supplier.
- Look for a supplier that covered your own uses with his registration.
- Provide supplier on request with Article 18(4) confirmation
- Use information for your own registration.
- Use information to prepare exposure scenarios to be included in CSR.
- Use information as a basis for a Downstream User report.

Afterwards, you may use this information for several purposes. First, you may inform your supplier to make your use(s) an identified use(s) to enable your supplier to cover your use(s) in his registration.

If your supplier already registered a certain substance this list may support you in checking the SDS/eSDS that your supplier provides you with.

It may also be useful to have this list available in case your supplier asks you for an Article 18(4) confirmation with the aim to register a certain substance as a transported isolated intermediate. You will be in the position to respond quickly whether this covers your use or not.

The list may also be useful for your Sourcing Department, if they are looking for further supplier(s) for a substance, as it will be clear whether the raw material needs to have a standard registration or whether it can be also bought from suppliers having registered a transported isolated intermediate only.

If it is necessary that your company registers a raw material in the role of an importer there will be already basic information available to make the decision concerning the type of registration that is required and to define quickly which data requirements are relevant.

For a Non-EU company the list may be useful for those raw materials that later on shall be forwarded to EU customers or when a supplier asks for an Article 18 (4) confirmation.

If it is clear that a substance can be registered as a transported isolated intermediate, the information given in the list below will be enough to fill in the corresponding Section 3.5 in the IUCLID5 Substance Data Set.

However, if there is a standard registration required, the information given in the list below may only be a starting point, as it may be necessary to develop exposure scenarios for inclusion in a CSR.

13.1.4
List of Raw Materials when Our Company is a Downstream User

It is useful to have a list containing all raw materials purchased from supplier(s) that deliver (pre-)registered substances for which your company benefits from being a Downstream User. This list may be helpful for determining the tasks to be done in your company to fulfill the obligations of a Downstream User.

Furthermore, it may be helpful for on-site inspections by national authorities as you can clearly state that your role under REACH is determined. If a substance is already registered by your supplier you may provide the national authorities on request with the registration number that covers the raw material purchased from a certain supplier.

If there is a discrepancy between the type of registration required for the use of a substance and the registration done by a certain supplier you may take further action – either purchase from another supplier or be aware of the fact that it may be necessary to do an own Downstream User report. An example of a list that may support you in identifying your obligations as a Downstream User can be found in Figure 13.6.

13.1 List of Used Raw Materials

Substance name	CAS	EC	Supplier	Supplier located in the EU/ Non-EU	OR for Non-EU supplier	Amounts of substance purchased by this supplier	Registration deadline of supplier or registration number	Type of registration supported by supplier	SDS/eSDS received from supplier
Substance 1	...	...	A	EU	–	15 t/a	2010	standard	✓
			B	Non-EU	...	20 t/a	2018	TII	
			C				01-....	standard	

Figure 13.6 List of raw materials for that our company is Downstream User.

TASKS to be done by a downstream user:

1. Provide your supplier on request with Article 18(4) confirmation.
2. Ask your supplier on a regularly basis whether he already has registered and if so ask for SDS/eSDS. If your supplier did not register yet, try to ensure that future demands concerning a certain substance will be covered and deliveries will be in legal compliance as the substance for example, had been pre-registered and the registration deadline of your supplier is based on tonnage band and properties of the substance at a later date.
3. If the substance as such is forwarded to your customers, ensure that you fulfill all your obligations to inform your customer accordingly about the hazards of a substance, risk management measures, etc. by providing a correct SDS/eSDS.
Furthermore, in the case of a transported isolated intermediate ask your customer(s) for an Article 18 (4) confirmation that has to be provided to national authorities on request.
4. Ensure that the substance purchased from your supplier is used at your site as foreseen in the SDS/eSDS.

13.1.5
Process after Receiving a SDS or an eSDS from Your Supplier

Under REACH, downstream users must not place on the market or use any substances that are not registered in accordance with REACH [2]. Although a downstream user in principle has no registration obligation for the substances purchased from sources that take care of the registration obligations, there are several other duties that downstream users have to fulfill up and down the supply chain. Most of these duties are connected with the safety data sheets that a downstream user receives from his supplier(s) for dangerous substances and preparations.

Some safety data sheets will have as an annex, a so-called exposure scenario [2]. This exposure scenario will give more specific information on how to use the substance or preparation safely [2].

After receipt of a safety data sheet a downstream user has the obligation to check it in depth, as he has the obligation to follow the instructions in the safety data sheets and if so in the exposure scenarios [2]. Therefore, the SDS/eSDS from the supplier needs to be examined and comparison with the uses of the downstream user is necessary. If a use of a downstream user is not covered by an exposure scenario, he may communicate with his supplier with the aim of having the use covered by an exposure scenario within the supplier's registration. If the registrant is not willing to cover the identified uses of the downstream user within his registration dossier, it needs to be listed as a use advised against. As a consequence, it may be necessary for the downstream user to develop their own chemical safety report.

A downstream user shall also contact his supplier(s), if he has new information on the hazard of a certain substance or preparation or if he believes that the risk management measures are not appropriate.

A downstream user may use a certain substance in a preparation or formulation or for the production of articles. In connection with these uses there may occur the obligation for the downstream user to provide his customers with certain information. He shall provide his customers with information on hazards, safe conditions of use and appropriate risk management advice concerning preparations. Concerning articles customers must be informed if the content of certain dangerous substances, which are candidates for authorization, exceed a concentration of 0.1% w/w in an article [2].

13.1.5.1 Four Key Steps in Checking Safety Data Sheets and Exposure Scenarios
There are four key steps that are essential in checking safety data sheets and exposure scenarios. Based on the outcome of the check of each of this steps there may be further action required. For a short overview see Figure 13.7.

13.1.5.2 Information to be Forwarded to Customers down the Supply Chain
If a downstream user produces preparations, he will have to forward to his customers within the safety data sheet that he develops, all the relevant information that

Task in checking SDS and exposure scenario	Further action
1. Read the description of use in the first part of the exposure scenario [2]	• If the description of use is very different from the way you use the product, you should contact your supplier and discuss it [2].
2. Compare the information given in the exposure scenario on how the substance or preparation may be used with your own use(s).	• If you use the substance or preparation in a way that leads to higher exposure, e.g. if you use it more often, in larger amounts or in a different way from that described, you may not comply with the exposure scenario and you should contact your supplier [2].
3. Compare the risk management measures specified in the exposure scenario to the way how you protect workers, consumers or the environment	• Decide if your measures are as, or even more efficient than those recommended in the exposure scenario [2]. • You should also inform your supplier if you think the risk management measures he recommends are inappropriate [2].
4. If your use of the substance or preparation differ from the exposure scenario, it may pose risks to your workers, consumers or the environment [2], therefore further steps must be defined.	There are a number of options [2]: • Contact your supplier and ask him to prepare an exposure scenario that fits your use conditions • change your working practices • assess in more detail if there is actually a risk or not • look for less hazardous substances or preparations that can be used alternatively

Figure 13.7 Key steps in checking safety data sheets and exposure scenarios and further action to be done.

he received himself from his supplier(s) in the safety data sheet and exposure scenario [2].

When forwarding a transported isolated intermediate, the customer shall be informed accordingly and shall be asked for an Article 18 (4) confirmation.

13.2
List of Substances that are Manufactured in Your Company

To find out whether your company has registration obligations concerning substances that are manufactured within your company, it is important to know whether a substance is manufactured in amounts of 1 t/a and above when your company is located within the EU. For a Non-EU company the amount manufactured per year is not relevant, but the amount of a substance that is imported into EU. As there are several options to achieve REACH compliance when a Non-EU manufacturer supplies EU customers, the process for identification of registration obligations needs a more detailed consideration for Non-EU manufacturers that may register via an Only Representative or make their EU customers act as importers that only when buying 1 t/a and above will have to register.

After having identified whether there are registration obligations, based on the properties of a particular substance the registration deadline for previously pre-registered substances can be determined.

In any case, a manufacturer will have to communicate with his customer(s) to achieve REACH compliance at the stage of defining what type of registration may be appropriate.

Within the next sections you will find some useful lists and short descriptions of tasks that will help you in defining registration obligations, registration deadlines and what type of registration may be required. Where it seems appropriate there is made a distinction between tasks for EU manufacturers and Non-EU manufacturers.

13.2.1
Identification of Registration Obligations

For a company located within the EU it is easy to find out when a substance needs to be registered as in general all substances manufactured in amounts of 1 t/a and above must be registered, if they cannot benefit from an exemption. Non-phase-in substances have to be registered before manufacture/import of 1 t/a and above, whereas for phase-in substances a company can benefit from later registration deadlines when a certain substance was pre-registered.

In Figure 13.8. you will find a table that may support you in defining the REACH registration obligations of your company, if your company is located within the EU.

The corresponding list in Figure 13.9 is appropriate for the needs of Non-EU manufacturers in determining REACH registration obligations.

13.2.2
Define the Registration Deadline for Substances Manufactured within Your Company Based on Their Properties and Consider Consequences if a Substance is Included in the SVHC Candidate List

Similar to the procedure described in Section 13.1.2 the properties of a substance as CMR and R50/53 may have an influence on the registration deadline for phase-in substances.

However, inclusion of a substance manufactured by your company in the candidate list for substances of very high concern and maybe inclusion in Annex XVII (subject to restrictions) or Annex XIV (subject to authorization) does not have an influence on the registration deadline. As it may be expected that this situation will have a far-reaching consequence on your business to be you may note the inclusion of a substance in the Candidate list and think through the consequences

13.2 List of Substances that are Manufactured in Your Company

Substance	CAS	EC	Phase-in or non-phase-in substance?	Is substance exempted from REACH?	Amount manufactured per calendar year (if so average of 3 last consecutive calendar years)	Amount imported from Non-EU per calendar year *)	Total amount to be considered	Registration obligation?	To Do
Hydrogen	...	...	...	yes	...	...	...	no	-
Product P	...	...	Phase-in	no	900 kg/a	200 kg/a	1.1 t/a	yes	Pre-registration and registration
Substance S	...	...	Phase-in	no	999 kg/a	-	999 kg/a	no	-
Substance B	...	...	Non-phase-in	no	5 t/a	-	5 t/a	yes	Inquiry and registration

Figure 13.8 List of substances manufactured by a company located in the EU and registration obligations.

*) if the Non-EU manufacturer does not take care of the registration obligations via an Only Representative

TASK: Define whether a substance needs to be registered by an EU manufacturer and define the necessary steps

1. List all substances manufactured by your company, if they are isolated (non-isolated intermediates must not be registered).
2. Check whether there are substances that are exempted from REACH registration obligations.
3. Add for all relevant substances the yearly manufactured amount. If the relevant amount is above 1 t/a (see also SVT list in Section 13.4.1) there are registration obligations.
4. If a substance is not only manufactured by your company but also imported from a Non-EU country your company may also have registration obligations as an importer under REACH and amounts manufactured and amounts imported must be added to determine whether you have registration obligations and also to determine the registration deadline.
5. For phase-in substances you may do a (late) pre-registration in due time and afterwards prepare a registration dossier in advance of the relevant registration deadline. If you manufacture a phase-in substance for the first time, please check whether late pre-registration is still possible or whether you are urged to do an Inquiry instead.

For non-phase-in substances you will have to submit an Inquiry Dossier to ECHA and afterwards you may submit the registration dossier.

Substance	CAS	EC	Is substance exempted from REACH?	EU-Customer(s)	Does customer take care of registration obligation?	Role of the EU customer	Amount supplied to EU customer(s) per calendar year (if so average of 3 last consecutive calendar years)	Amount to be considered for registration via our OR?	TO DO
Product P	...		no	Customer 1	yes	I	18 t/a	–	
				Customer 2	no	DU	5 t/a	yes	
				Customer 3	?	DU	85 t/a	yes	
				Customer 4	no	DU	5 t/a	yes	
Product P	Total amount to be registered by our company via OR: 95 t/a, which means tonnage band 10 to 100 t/a								

Figure 13.9 List of substances manufactured by a Non-EU manufacturer and determining REACH registration obligations.

TASK: Define whether a substance manufactured by a Non-EU manufacturer needs to be registered

1. List all substances manufactured by your company, if they are sold to EU customers.
2. Check whether there are substances that are exempted from REACH registration obligations.
3. Mark all customer(s) that will take care of the registration obligations in the role of an importer under REACH.
4. Add for all relevant substances the amount sold to each EU customer per calendar year (if so each calendar year since 2007 – see also SVT list in Section 13.4.1).
5. When there are customers that purchase less than 1 t/a and they do not purchase the same substance from any other source that did not (pre-)register there may be no registration obligation for these customers.
6. For all other EU-customers you may have, the amounts delivered to them per calendar year shall be summed up to define the relevant tonnage band that may be registered by your company via an Only Representative within the EU.
7. Based on whether the substance is a phase-in substance that has been pre-registered in due time or a non-phase-in substance the steps on the way to registration may be defined. Inquiry and afterwards registration for non-phase-in substances and phase-in substances that were not (late) pre-registered in due time. (Late) pre-registration and at a later stage registration have to be done for phase-in substances that were pre-registered in due time.

for the future. The consequences for your company as a manufacturer will be different from the consequences that a downstream user will have to face.

When a certain substance is manufactured at your site under strictly controlled conditions and can be registered as a transported isolated intermediate, you may not fear the consequences from an eventual inclusion of such a substance into Annex XIV as transported isolated intermediates will not be subject of an authorization. If you manufacture a substance that needs a standard registration for example, because of the use as a solvent, your company may directly be concerned when this substance will be included in Annex XIV. In such a case your company may balance the advantages and disadvantages of manufacturing that substance in future. Maybe you will make the decision to cease manufacture of a certain substance at the moment when it becomes subject to authorization, because the efforts and costs connected to the application for authorization bear a high risk compared to the profit you actually have because of selling this substance.

A table that may be used to list the substances manufactured within your company and their relevant properties to determine the registration deadline can be found in Figure 13.10.

The list may be used by EU manufacturers (considering the volumes manufactured per calendar year) and Non-EU manufacturers that wish to register a certain substance via an Only Representative in the EU. For Non-EU manufacturers only amounts supplied to EU customers per calendar year must be considered.

13.2.3
Uses at the Company's Own Site and Identified Uses of the Customers

If a company is located within the EU they may manufacture and isolate a substance that is afterwards used at their own site for the synthesis of another substance. If manufacturing and use can be done under strictly controlled conditions, the company may benefit from the cheap version of a registration as an on-site isolated intermediate.

If that intermediate is sold to customers and therefore transported to another site a registration as a transported isolated intermediate may be appropriate.

For all cases in which a substance is used not in accordance with the strictly controlled conditions or not for the synthesis of another substance a standard registration is required.

To define which type of registration will be required for a certain substance manufactured within your company and may be sold to customer(s), you may have a list with all substances manufactured at your site. Then add all the uses of your own company and also the uses of all your customers. If you do not know the uses of your customers yet and they did not provide you with a list of their uses to make them identified uses, you may ask them to provide you with some details.

An example list for identification of the relevant uses of a substance can be found in Figure 13.11.

Substance	CAS	EC	R50/53	CMR	Included in SVHC candidate list *)	Tonnage band			Registration deadline
						1–10 t/a	10–100 t/a	100–1000 t/a	>1000 t/a	
Substance 1	...	...	x			x				31.05.2018
Substance 2	...	...	x					x		30.11.2010
Substance 3	...	...		x		x				30.11.2010
Substance 4	...	...	–	–				x		31.05.2013
...										
Substance x	...	...	–	–			x			31.05.2018

Figure 13.10 Properties of substances manufactured by your company and corresponding registration deadline for pre-registered substances.
*) it could also be checked whether a certain substance is subject to restrictions or subject to authorization by checking the relevant Annnexes of the REACH regulation.
Remark: Inclusion in SVHC Candidate list does not directly influence the registration deadline, but may be a useful information to develop business strategies in due time.

13.2 List of Substances that are Manufactured in Your Company | 223

Substance	CAS	EC	Customer	Type of registration required			Uses described by Use descriptors (1 line for each use)					To Do
				Standard registration	TII	OII	ERC	PROC	PC	Technical function of the Substance	SU	
Substance 1	...	...	Customer 1	x			...	...	...	...	...	
			Own company and Customer 2		x		6a	1	19	Intermediate	8, 9	Registration as TII
			Customer 3	x				4	19	Intermediate, but not under scc		Standard registration required, provide eSDS
			Customer 3	x						Catalyst		
			Customer 4	x						Monomer for manufacturing of a polymer		
			Customer 1 and 2	x						solvent		

Figure 13.11 Identify uses of a certain substance within your company and at your customers.

Remark: This list in principle may be used by companies located within the EU and also by Non-EU manufacturers that will take care of the registration obligations via an Only Representative, but in case of a Non-EU manufacturer the column OII does not apply.

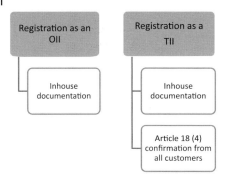

Figure 13.12 Documentation for on-site isolated intermediates and transported isolated intermediates.

Afterwards you may decide which type of registration will be appropriate.

Directly connected to the type of registration and tonnage band there will be the decision whether you will have to provide a SDS or eSDS to your customers after having successfully registered a certain substance.

13.3
Documentation Concerning Manufacturing Process of OIIs and TIIs and Documentation of the Correct Use of TIIs by Customers

In case of on-site inspections the manufacturer has to proof that the substance is handled at his site in accordance with the strictly controlled conditions by having available for example, in-house documentation. For transported isolated intermediates there are also confirmations from the customers required that they use the substance in accordance with Article 18 (4) of the REACH regulation [1], see Figure 13.12.

13.3.1
In-house Documentation

The in-house documentation concerning handling of a substance as on-site isolated intermediate or transported isolated intermediate can be done by using the forms that were presented in Chapter 5 and especially in Figure 5.6.

13.3.2
Confirmation from Downstream Users Concerning Art. 18(4) and Further Exemptions

Examples for Article 18 (4) confirmations requested by customers were already presented in Section 5.3 of this book. To ensure REACH compliance within your company it is recommended to ensure based on IT-tools used for the order management that an order within your company cannot be processed and the substance cannot be supplied without having the Article 18 (4) confirmation from the corresponding customer.

13.4
Substance Volume Tracking

Substance Volume Tracking (SVT) is a task that shall be done on a regularly basis at least once per year with the aim to find out whether a pre-registration or registration needs to be updated because of a change of tonnage band (increase or decrease of tonnage band). Cessation or re-start of manufacturing of a certain substance may also demand further action. The point of view is different for EU and Non-EU manufacturers and needs to be considered.

13.4.1
Substance Volume Tracking for EU Manufacturer

For the SVT done by a company located within the EU there has in general the whole amount of a substance manufactured within a calendar year to be considered. For import of substances the imported volume per calendar year has to be considered. If a certain substance is manufactured within the company and also imported by the same company the relevant amount is the total of manufactured and imported volumes.

A list that may support a company located within the EU by doing the SVT can be found in Figure 13.13.

13.4.2
Substance Volume Tracking for a Non-EU Manufacturer

SVT for a Non-EU manufacturer may be more demanding than for an EU manufacturer, because of the different point of view. The yearly manufactured volumes of a certain substance need not be considered, but the amount of a substance that is imported into EU by the Non-EU manufacturer either directly or via a Non-EU distributor that was allowed to refer to the Non-EU manufacturers registration (done via an Only Representative within the community) needs consideration.

Difficulty may also arise in a situation in which a Non-EU manufacturer via an Only Representative takes care of the registration obligations of parts of the volumes imported into EU, but there is also one or several customers that registered on their own and therefore act as importer(s). In this case, the whole amount of a certain substance that is delivered to EU customers need not be considered, but only the amount that is covered by the Only Representative's registration.

An example of a list to support SVT of a Non-EU manufacturer is presented in Figure 13.14.

Substance	CAS	EC	Amount manufactured/imported							Average of 3 last consecutive calendar years	Tonnage band (pre-)registered	Tonnage band compared to (pre-)registration	To Do
			2007	2008	2009	2010	2011	2012	2013				
			…	…	…	…	…	…	…	9 t/a	1 to 10 t/a registered	still OK	–
			…	…	…	…	…	…	…	142 t/a	10 to 100 t/a registered	increased	Update of registration
												decreased	
											1 to 10 t/a pre-reg	cessation	set pre-registration to "inactive"

Figure 13.13 SVT list for a company located within the EU.

TASKS:
1. List all substances manufactured/imported per calendar year for which you have registration obligations.
2. Note the amounts manufactured/imported since 2007. If the substance was manufactured/imported in each of the years the relevant actual tonnage band may be calculated on the average of the three last consecutive calendar years.
3. Compare the relevant calculated tonnage band with the tonnage band indicated in the pre-registration or registration.
4. If the calculated tonnage band deviates from the volumes given in the pre-registration or registration, please take further action according to the actual situation.

13.4 Substance Volume Tracking | 227

Substance	CAS	EC	EU customer (role)	Amount imported into the EU per calendar year							Average of 3 last consecutive calendar years	Tonnage band (pre-)registered for my company	To Do
				2007	2008	2009	2010	2011	2012	2013			
Substance 1	...	...	Customer 1 (DU)	...	...	...	...	...	...	...	23 t/a		
			Customer 2 (I)								not relevant		
			Customer 3 (DU)								87 t/a		
Substance 1	**Total to be considered for each calendar year:**										To be considered: 110 t/a	10 to 100 t/a	Update, increase of tonnage band

Figure 13.14 SVT list for a company located outside of the EU that has an Only Representative.
TASKS:
1. List all substances supplied to EU customers and list all relevant EU customers.
2. Define the role of each EU customer (DU or I).
3. Add the relevant amounts supplied to each customer in the relevant calendar years since 2007.
4. Sum up the volumes to be considered by your company (taking care for the registration obligations via Only Representative).
5. Check whether the relevant calculated tonnage at present still fits the tonnage band given in pre-registration or in the registration dossier.
6. In case of deviations take further action according to the situation.

13.5
Examples and Exercises

1. What is the most important difference in doing Substance Volume Tracking for a company located within the EU compared to the Substance Volume Tracking that may be done at a Non-EU manufacturer's site?

2. What duties does a downstream user have after receipt of a SDS/eSDS by his supplier(s)?

3. Is there any situation in which it could be important for a Non-EU company to purchase raw material that is (pre-)registered?

4. A company located in Switzerland purchases a certain substance used as a raw material from a Chinese Distributor. This substance will be used for the synthesis of another substance P at a Swiss site.

    a) Which registration obligations will the Swiss company have concerning the substance purchased from the Chinese distributor?

    b) Which role under REACH will the Swiss company have when the product P is supplied as a 50% solution in cyclohexane to

    - a customer located in the USA;
    - a customer in Switzerland;
    - an Austrian company?

    c) Who will have registration obligations in the cases described in b)?

5. What sort of risks has a company to face, if they purchase a raw material that is included in the SVHC candidate list?

References

1 Regulation (EC) No 1907/2006 of the European Parliament and of the Council of 18 December 2006, concerning the Registration, Evaluation, Authorisation and Restriction of Chemicals (REACH), establishing a European Chemicals Agency, amending Directive 1999/45/EC and repealing Council Regulation (EEC) No 793/93 and Commission Regulation (EC) No 1488/94 as well as Council Directive 76/769/EEC and Commission Directives 91/155/EEC, 93/67/EEC, 93/105/EC and 2000/21/EC.

2 Guidance for downstream users, January 2008, published by ECHA.

14
Communication in the Supply Chain

14.1
Communication Obligations According to the REACH Regulation

REACH [1] requires a lot of communication within pre-SIEFs and SIEFs, but there are also a lot of tasks that require communication within the Supply Chain. Sometimes, as a consequence of the outcome of previously communication in the Supply Chain it is also required that an actor in the Supply Chain who is not a registrant will have to communicate with authorities.

The most important Articles within the REACH regulation [1] concerning communication obligations within the Supply Chain are listed in Figure 14.1. In connection with the corresponding Article you will find a hint as to who in the Supply Chain may be concerned and therefore will have to fulfill communication obligations. For an overview on possible actors within a supply chain see Figure 14.2. As a single manufacturer or supplier may be the supplier for more than only one company, down the supply chain in every step there may be a multiplication of involved actors. This is indicated by the spreading arrow in Figure 14.2.

There is the general obligation to keep information [2] that concerns all actors in the Supply Chain at any time in accordance with Article 36 (1) [3]: "Each manufacturer, importer, downstream user and distributor shall assemble and keep available all the information he requires to carry out his duties under this Regulation for a period of at least 10 years after he last manufactured, imported, supplied or used the substance or preparation." On request this information shall be made available to any competent authority of the Member state in which he is established or to ECHA.

Part of the REACH Regulation [1]		Task(s) for
Title II, Article 8	Only Representative of a non-Community manufacturer	
Article 8 (1)		Non-EU manufacturer
Article 8 (2), (3)		Only Representative
Title IV	INFORMATION IN THE SUPPLY CHAIN	
Article 31	Requirements For Safety Data Sheets	Manufacturer and Downstream Suppliers
Article 32	Duty to communicate information down the supply chain for substances on their own or in preparations for which a safety data sheet is not required	Manufacturer and Downstream Suppliers
Article 33	Duty to communicate information on substances in articles	Supplier
Article 34	Duty to communicate information on substances and preparations up the supply chain	Any actor in the supply chain
Article 35	Access to information for workers	Employers (Manufacturers, Importers, Downstream Users)
Article 36	Obligation to keep information	All actors in the Supply Chain: Manufacturers, Importers, Downstream Users, Distributors
Title V	DOWNSTREAM USERS	
Article 37	Downstream user chemical safety assessment and duty to identify, apply and recommend risk reduction measures	Downstream Users, Distributors in connection with Manufacturers
Article 38	Obligations for downstream users to report information	Downstream Users
Article 39	Application of downstream user obligations	Downstream Users
Title VII, Chapter 3	Authorisations in the supply chain	
Article 65	Obligation of holders of authorisation	Holders of Authorisation, Downstream Users
Article 66	Downstream users	Downstream User

Figure 14.1 Articles of the REACH regulation [1] that are connected to communication obligations within the Supply Chain.

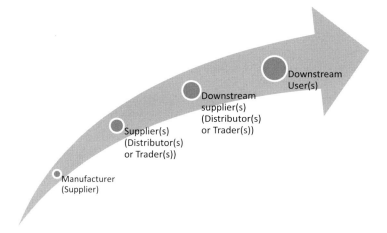

Figure 14.2 Actors in the Supply Chain.

14.2 Communication to be Done by Suppliers

Suppliers located within the EU	• EU Manufacturer • EU Trader/Distributor (Downstream User or Importer)
Suppliers outside of the EU	• see Section 14.3 Communication to be done by Non-EU manufacturers and Section 14.4 Communication to be done by Non-EU Distributors or Non-EU Traders

Within this section we will focus on tasks to be done by EU suppliers. Tasks of suppliers that are established outside of the EU (Non-EU suppliers) are considered in Section 14.3 by assuming that a Non-EU manufacturer is acting as a supplier. This Non-EU supplier may take over the registration obligations via an Only Representative. The Only Representative in principle fulfills the registration obligations of an importer. The communication obligations of an importer under REACH will be similar to those of an EU-manufacturer. Communication to be done by Non-EU Traders or a Non-EU distributor will be covered in Section 14.4.

14.2.1 Communication from Supplier to EU Customers

Duties and tasks that may be relevant in the relationship between EU suppliers and their EU customers are listed in Figure 14.3. In the second column you will find also hints concerning relevant parts within the REACH regulation and if so it is referred to sample documents as sample letters that are included in this book and that may support you as a supplier in communicating with your EU customers.

Duties and tasks	Hints
• Information to customers that substances supplied to him have been pre-registered (either by supplier itself or any actor up the supply chain)	→Sample Letter in Figure 14.4
• Ask for uses of customers	→Sample Letter in Chapter 5, Figure 5.1
• If registration as a transported isolated intermediate is intended to be done, ask customer for Article 18 (4) [4] confirmation	→Sample Letter in Chapter 5, Figure 5.3
• provide SDS/eSDS to customers and if so also necessary updates	→ REACH Article 31 [5]
Alternatively: Duty to communicate information down the supply chain for substances on their own or in preparations for which a safety data sheet is not required	→REACH Article 32 [6]
• Duty to communicate information on substances in articles	→ REACH Article 33 [7]

Figure 14.3 Communication from EU Supplier to EU customer.

REACH – Information on substances that we or our upstream suppliers already pre-registered/late pre-registered

Dear valued customer,

We *[company name]* can confirm that our company pre-registered the

substance *[substance name]*, CAS *[CAS number]*, EINECS *[EINECS number]*

that your company purchases from us. Therefore your company will be a Downstream User without any registration obligations.
We intend/Our upstream supplier intends also to register the above mentioned substance. To find out which type of registration will be appropriate and which uses shall be covered, we will sent a questionnaire to our customers to identify their uses in due time.

Kind regards,
[company name]

Figure 14.4 Sample letter to inform a customer that a substance supplied to him is (late) pre-registered.

14.2.2
Communication from Supplier to Non-EU Customers

As we assume within this section that a Supplier is located within the EU, whereas the situation for a Non-EU Supplier who may take care of registration obligations is covered by the role of a Non-EU manufacturer, there will be REACH registration obligations that are either covered by the EU supplier himself or by any actor up

Duties and tasks	Hints
• Ask for uses of customers	→Sample Letter in Chapter 5, Figure 5.1
• If registration as a transported isolated intermediate is intended to be done, ask customer for Article 18 (4) [4] confirmation	→Sample letter see Chapter 5, Figure 5.3
• If customer is not willing to support his supplier as REACH is not applicable at his site, the supplier may alternatively send letter to this customer demanding that the substance is handled accordingly	→Sample letter see Chapter 5, Figure 5.4
• provide SDS/eSDS to customers and if so also necessary updates	→ REACH Article 31 [5]
• Alternatively: Duty to communicate information down the supply chain for substances on their own or in preparations for which a safety data sheet is not required	→REACH Article 32 [6]

Figure 14.5 Communication from an EU-supplier to Non-EU customers.

the Supply Chain. The Non-EU customers in general do not have any obligation to cooperate with their EU suppliers in terms of REACH, but they may be willing to support their suppliers especially in cases when the registration costs can kept at a lower level, which will also be for the benefit of a Non-EU customer when the price for the product that he purchases will not increase.

Some cases in which an EU supplier will communicate with a Non-EU customer to fulfill his obligations under REACH are listed in Figure 14.5.

14.2.3
Information for Workers

A supplier located within the EU has not only communication obligations with external parties, but also the duty to enable access to certain information for the workers within his company.

Article 35 [8] of the REACH regulation [1] demands "Workers and their representatives shall be granted access by their employer to the information provided in accordance with Articles 31 [5] and 32 [6] in relation to substances or preparations that they use or may be exposed to in the course of their work." Figure 14.6 shows which information an EU-supplier shall make available to his workers.

It is clear that every employer will have an interest in the protection of his workers and also will be interested in avoiding any accident within his company, therefore any employer that feels responsible for his workers will grant them access to all information that is available from SDS/eSDS and if available also from further sources.

Duties and tasks	Hints
• Give workers access to information provided in accordance with Article 31 and Article 32	→REACH Article 35 [8]
• provide SDS/eSDS to workers and if so also necessary updates	→ REACH Article 31 [5]
Alternatively: Duty to communicate information for substances on their own or in preparations for which a safety data sheet is not required	→REACH Article 32 [6]

Figure 14.6 Information that an EU supplier shall forward to his workers.

As it is not determined by the REACH regulation how this should be done, an employer may feel free to choose a pragmatic way to fulfill his obligation. There may be companies that provide their workers with the Safety Data Sheets as a paper version, others may give additional information within a short lecture or even check by IT-tool-supported tests whether the workers have the knowledge to handle a certain substance safely.

14.2.4
Communication with Upstream Supplier

If a supplier communicates with an upstream supplier, he will act either in the role of a Downstream User (more exactly: Downstream Supplier) or in the role of an Importer. He may act as a Downstream User, if the upstream supplier either is located within the EU or when he is a Non-EU manufacturer who takes care of the registration obligations via an Only Representative. Only if the upstream supplier is a Non-EU supplier who is not willing or is not allowed to take care of the registration obligations, will the supplier act in the role of an importer with own registration obligations.

First, it will be necessary to find out which role the Supplier will have, therefore he may ask the upstream supplier for the corresponding information. A sample letter and questionnaire was provided in Chapter 2, Figure 2.2 of this book.

Further communication that may be done in the role of the Downstream User can be found in Section 14.5. Further communication in connection with a role as an importer under REACH will be similar to the communication to be done by an EU manufacturer, but does normally concern actors down the supply chain.

14.2.5
Communication with Authorities

It is clear that any supplier that is an EU manufacturer will communicate with ECHA as he may highly likely have registration obligations. In general, an EU manufacturer will also be used to communicate with national authorities concerning a number of issues and national authorities will visit an EU-manufacturer with the aim to do on-site inspections to check whether an EU-manufacturer fulfills his obligations under REACH. The same in principle can be applied to importers.

The type of relationship and therefore also the type of communication may be different for Suppliers that are acting as Traders or Distributors. Under REACH they may have to cope with on-site inspections done by national authorities as any other Downstream User, but as they do not really use a certain substance it is more appropriate to call them Downstream Suppliers.

A Downstream Supplier has similar duties as a Downstream User, but there is one big difference. The Downstream Supplier normally will not be urged to communicate with ECHA, whereas a Downstream User under certain circumstances may have to communicate with ECHA for example, when there is an obligation to compile a Downstream User Report. This will be explained in Section 14.5.5.

14.2.6
Further Communication Obligations for Suppliers in the Context of Authorisation

Obligations in the supply chain resulting from authorisation are defined in Article 65 [9] and Article 66 [10] of the REACH regulation [1].

Figure 14.7 shows the communication obligations that a supplier may have towards his customers and also the obligation of supplier and customer to communicate with ECHA under certain circumstances.

Supplier to customers	• Holders of an authorization and Downstream users shall include the authorization number on the label before they place on the market a particular substance or preparation [9]
	• Shall be done without delay once the authorization number has been made publicly available [9]

Supplier/customer(s) to notify ECHA	• If supplier is also Downstream user he shall notify ECHA within three months of the first supply of the substance [10] • Any Downstream User (in general customers of the supplier) shall notify ECHA within three months of the first supply of the substance [10] → Supplier shall inform his customers as soon as possible

Figure 14.7 Communication of an EU supplier in the context of authorisation.

14.3
Communication to be Done by Non-EU Manufacturers

Non-EU Manufacturer	• Any Manufacturer whose site is located outside of the EEA

14.3.1
Communication from Non-EU Manufacturer to his Only Representative

A Non-EU Manufacturer may take care of the registration obligations under REACH by appointing an Only Representative with the intention to make his EU customers benefit from being Downstream Users. As a consequence of appointing an Only Representative the Non-EU Manufacturer has to take care that his Only Representative is provided with all necessary information not only to compile the registration dossier, but also information about ongoing negotiations that is required by the Only Representative for on-site inspections by national authorities or on request from national authorities or even ECHA.

In Figure 14.8 you will find issues that may be the subject of communication from a Non-EU manufacturer to his Only Representative.

14.3.2
Communication from Non-EU Manufacturer to EU Customer

In general, a Non-EU manufacturer does not have any obligation under REACH and therefore is not required to do any communication about this issue with his EU customers without request. However, as soon as a Non-EU manufacturer makes the decision to take care of registration obligations via an Only Representative there will be further communication obligations to be managed by the Non-EU manufacturer.

Situations in which a Non-EU manufacturer will communicate with his EU customers concerning REACH matters are listed in Figure 14.9. In the second column you will find hints on the corresponding part within the REACH regulation and also a connection to sample letters that are included in this book.

14.3.3
Communication from Non-EU Manufacturer to Non-EU Customers

It is difficult to give any recommendations concerning the necessity of doing any communication between Non-EU manufacturers and Non-EU customers concerning REACH topics. In principle, this sort of relationship is not under the scope of REACH. However, there may be REACH issues to be discussed in the case of indirect imports into the EU, while at the same time the competition law has to be respected. The scheme in Figure 14.11 shows how a substance manufactured by a Non-EU manufacturer may be distributed to customers within the EU and outside of the EU. The Non-EU manufacturer will be aware of the volumes directly sold to EU customers (direct imports), but may not be aware of the volumes that may be brought to the EU by one or more of his Non-EU customers at a later stage.

A Non-EU manufacturer may allow all Non-EU customers to forward a certain substance to EU sites or EU customers as he registered via an Only Representative. In this case, there needs to be found a solution of how the OR may know the exact

Duties and tasks	Hints
• Appointment letter in accordance with Article 8 (1) [20]	→see sample letter in Chapter 6, Figure 6.13 – this letter may be used by the OR as attachment in the registration dossier. This sort of Certificate may be sufficient within a company group to enable the OR to fulfill the duties as against ECHA. When a Non-EU Manufacturer instructs an independent consulting company to act as his Only Representative, additional contractual arrangements may be necessary to determine rules for their cooperation including e.g. payment conditions.
• Provide OR with information that an OR shall have in accordance with Article 8 (2) [21]	
• quantities imported into EU	→Result of Substance Volume Tracking (see list in Chapter 13, Figure 13.14) may be delivered to an Only Representative on a regular basis. Within a company group it may be more convenient to have this information available via an IT-tool, e.g. SAP to which the EU legal entity that acts as an Only Representative for a Non-EU entity and the Non-EU manufacturer both have access
• list of customers sold to	→Duty may be fulfilled by providing a list to the Only Representative. Within a company group it may be more convenient to have this information available via an IT-tool, e.g. SAP to which the EU legal entity that acts as an Only Representative for a Non-EU entity and the Non-EU manufacturer both have access
• information on the supply of the latest update of the safety data sheet referred to in Article 31 [5]	→Duty may be fulfilled by providing a list to the Only Representative. Within a company group it may be more convenient to have this information available via an IT-tool, e.g. SAP to which the EU legal entity that acts as an Only Representative for a Non-EU entity and the Non-EU manufacturer both have access

Figure 14.8 Communication from Non-EU manufacturer to his Only Representative.

Duties and tasks	Hints
• Info that an OR has been appointed and therefore direct importers in EU will be Downstream users Alternatively if the Non-EU manufacturer does not want to take care of the registration obligations via an OR, it is recommended to inform the customers accordingly	→Article 8(3) [13] →Sample letter in Figure 14.10
• Ask EU customers for their uses to be covered within the registration dossier and if so within the CSR	→ Sample letter in Chapter 5, Figure 5.1
• When a registration as a transported isolated intermediate shall be done, ask for Article 18 (4) [4] confirmation	→ Sample letter in Chapter 5, Figure 5.3
• Provide SDS/eSDS including registration number and if so updates. SDS/eSDS may include contact details and address of OR to ensure that customers are informed about the appointment of the OR as requested by Article 8 (3) [13] • Alternatively: Duty to communicate information down the supply chain for substances on their own or in preparations for which a safety data sheet is not required	→ REACH Article 31 [5] → Article 8 (3) [13] → REACH Article 32 [6]

Figure 14.9 Communication from Non-EU manufacturer to his EU customers.

volumes and also the identity of the importers in the EU that shall benefit from being Downstream users.

Actually, it would be necessary that Non-EU customers that forward a substance into the EU provide the OR of the Non-EU manufacturer with the figures and seek for acceptance, as otherwise the importer in the EU has to register the substance in the role of an importer.

When a globally acting company that bought a certain substance from a Non-EU manufacturer for a Non-EU site, decides later on to forward the volumes of this substance as a whole or in parts to another site within the EU, it may seek an agreement with the Non-EU manufacturer to have this import covered under the registration of the Non-EU manufacturer's Only Representative. The Non-EU manufacturer may agree if there will no tonnage band increase to be expected and all actors in the supply chain will benefit from having less work.

Much more critical will be the situation when a Non-EU company purchased a certain substance from a Non-EU manufacturer and intends to distribute this

14.3 Communication to be Done by Non-EU Manufacturers

> **REACH – Appointment of Only Representative for *[name of NON-EU manufacturer´s company]***
>
> Dear valued customer,
>
> Our Company *[name of NON-EU manufacturer´s company]* in accordance with Article 8 of the Regulation (EC) No 1907/2006 of the European Parliament and of the Council of 18 December 2006 (REACh regulation) has appointed *[name and address of Only Representative company]* to act as our Only Representative for the following substances that are purchased from your company:
>
> [substance name], CAS [CAS number], EC [EINECS number]
> [substance name], CAS [CAS number], EC [EINECS number]
> [substance name], CAS [CAS number], EC [EINECS number]
> [substance name], CAS [CAS number], EC [EINECS number]
> [substance name], CAS [CAS number], EC [EINECS number]
>
> If you supply these substances from *[name of NON-EU manufacturer´s company]* you will benefit from being a Downstream User without having any registration obligations.
>
> Best regards,
> *[name of NON-EU manufacturer´s company]*

Figure 14.10 Sample letter to inform customers about Appointment of an Only Representative.

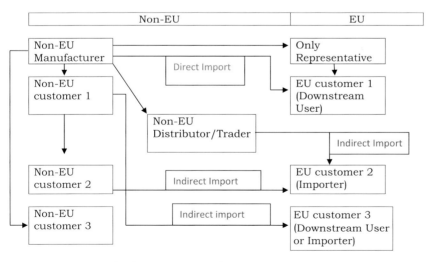

Figure 14.11 Direct and indirect imports into the EU.

volume to one or several customers within the EU. Concerning REACH it may be no problem to cover these amounts with the registration that a Non-EU manufacturer did via his Only Representative especially when the registration was done in the highest tonnage band. But there will be some traps, as REACH requires that the Only Representative can provide authorities with the exact figures of the imported amounts per calendar year and also the identity of the importers, whereas it may be a breach of the competition law when the Non-EU manufacturer is informed by his Non-EU customer about the identity of his customers and his business relationships.

However, if contact details of the Only Representative are forwarded in the Supply Chain it may be possible for the last distributor acting in the supply chain outside of the EU or the importers within the EU to cooperate with the Only Representative without disclosing confidential business information up the supply chain. In reality, this may cause problems when the Only Representative of the Non-EU manufacturer is not independent of the Non-EU manufacturer for example, a sister company established within the EU. If the Only Representative is acting independent of the Non-EU manufacturer for example, a consultant company this problem may be avoided. Then, there still remains the problem that an Only Representative will be provided with the figures for import after the import has been done. If the Non-EU manufacturer produces higher amounts of a certain substance than are covered within the registration there may occur further problems. If the figures require an update because of an increase in tonnage band, it cannot be traced back whether this is justified and there is no chance for the Non-EU manufacturer to avoid a tonnage band increase if he was willing to cover all indirect imports.

A Non-EU Manufacturer who is willing to cover not only direct imports of a certain substance, but is also willing to support his Non-EU customers that may forward this substance to their EU customers, will have to think through the situation with due diligence.

If the Only Representative of the requested Non-EU manufacturer already did a registration in the highest tonnage band and the type of registration is a standard registration there may be good reasons to also offer Non-EU customers to forward a certain substance to their EU customers, whereas for transported isolated intermediates there will occur a further problem in so far as the registrant must have an Article 18 (4) confirmation from all EU customers that intend to benefit from being Downstream Users within this supply chain. In Figure 14.12 there are listed situations when a Non-EU manufacturer may also inform his Non-EU customers concerning REACH issues. In Figure 14.13 you will find a sample letter that may be used to inform Non-EU customers how they may act if they intend to forward to EU customers a certain product that was supplied from the Non-EU manufacturer.

In all cases where the Non-EU manufacturer did a registration in a lower tonnage band than the highest one and also in cases where the registration was done for a transported isolated intermediate only, it must be recommended that the Non-EU manufacturer does not allow his Non-EU customers to benefit from

Duties and tasks	Hints
• Decision whether also indirect imports shall be covered with the registration done by the Non-EU manufacturer via Only Representative	→ If so, the Only Representative and also the Non-EU customers shall be informed accordingly
• Info of Non-EU manufacturer to Non-EU customers whether they may benefit from having covered amounts to be forwarded to EU customers by the registration of the Non-EU manufacturer that registered via an Only Representative	→ Sample letter in Figure 14.13 Important remark: When a transported isolated intermediate is to be covered, there must be found a solution as to how to ensure that the Article 18 (4) confirmations will be passed on in the supply chain and the letter needs to be adapted accordingly.
• Different versions of SDS/eSDS (not required if also all indirect imports shall be covered in the registration of the Non-EU manufacturer via Only Representative)	• version for EU customers including registration number • version for Non-EU customers without registration number (also applicable in case an EU customer registered on his own)

Figure 14.12 Communication from Non-EU manufacturer to Non-EU customers.

a sort of free ride when they forward a certain substance to EU customers at a later stage. If the Non-EU manufacturer intends to prevent his Non-EU customers from using the registration number without his knowledge, a further consequence will be that there are at least two versions of SDS: a version including the registration number that is distributed to EU customers that are doing direct imports and another version without any hint of the registration and registration number for all Non-EU customers.

14.3.4
Communication from Non-EU Manufacturers to Their Suppliers

Communication from a Non-EU manufacturer to his suppliers concerning REACH topics has to be made in only two situations. It is recommended to support your supplier whenever he requests an Article 18(4) confirmation from your company and you can confirm that a certain substance is handled at your site accordingly. A Non-EU manufacturer is not obliged to sign such a document, but may benefit directly, when his supplier can benefit from a cheaper type of registration.

The second situation in which communication from a Non-EU manufacturer to his supplier is required concerning REACH issues, is any situation in which the Non-EU manufacturer purchases a substance that at a later stage will be forwarded

Dear valued customer,

Our Company **[name of NON-EU manufacturer's company]** in accordance with Article 8 of the Regulation (EC) No 1907/2006 of the European Parliament and of the Council of 18 December 2006 (REACh regulation) has appointed **[name and address of Only Representative company]** to act as our Only Representative for the following substances purchased by your company:

[substance name], CAS [CAS number], EC [EINECS number]
[substance name], CAS [CAS number], EC [EINECS number]
[substance name], CAS [CAS number], EC [EINECS number]
[substance name], CAS [CAS number], EC [EINECS number]
[substance name], CAS [CAS number], EC [EINECS number]

If you supply these substances from **[name of NON-EU manufacturer's company]** and you intend to forward the substance to any customer located within EU, we can offer that these importers can benefit from being a Downstream User without having any registration obligations, if our Only Representative **[name and address of Only Representative company]** is informed accordingly.

The Only Representative **[name and address of Only Representative company]** has an ongoing obligation to provide authorities on request with the exact figures of imported volumes of substances that were registered by them on behalf of the Non-EU manufacturer. If any importer within the supply chain desires to be regarded as a Downstream User of **[name and address of Only Representative company]** they will have the obligation to inform **[name and address of Only Representative company]** about the imported volumes per calendar year.
If an EU importer does not inform **[name and address of Only Representative company]** about the amounts imported into EU, the importer may still have the registration obligation as foreseen by the REACH regulation.

We would like to remind you that you are responsible for providing details regarding your EU customers (name and address of the company, contact person, email, phone number) and the volumes exported to each EU customer per calendar year.

If you intend to forward these substances from **[name of NON-EU manufacturer's company]** to any other Downstream User or Downstream Supplier outside of EU, please pass on the information concerning our Only Representative.

Best regards,
[name of NON-EU manufacturer's company]

Figure 14.13 Sample letter to inform Non-EU customers what to do in case they intend to forward a certain substance to their EU customers by using the registration of the Non-EU manufacturer's Only Representative.

Duties and tasks	Hints
• Provide EU suppliers on request with Article 18(4) [4] confirmation	Supplier will be enabled to benefit from a less expensive type of registration, that may also be for your benefit, although REACH is not applicable outside of the EU and actually no European authority is in the position to check the handling of the substance at a Non-EU site
• If a certain substance shall be forwarded to the EU later on (Re-import), e.g. a solvent used to dissolve any substance or a substance used to prepare a formulation, ask your supplier whether the substance is (pre-) registered and check whether uses of the customer will also be covered with the registration of the supplier.	→Sample letter in Chapter 2, Figure 2.2 may be adapted for the Non-EU manufacturer.

Figure 14.14 Communication from Non-EU manufacturers to their suppliers.

to EU customers. When the substance purchased by the Non-EU manufacturer is from a registered source and the substance itself is not transformed into another substance it may be possible to avoid further registration obligations when it can be assured that the substance is re-imported as such. This will be for the benefit of the party that is re-importing such a substance, as there will be no further registration obligation any longer. Duties and tasks in regard to communication from Non-EU manufacturers to their suppliers can be found in Figure 14.14.

14.4
Communication to be Done by Non-EU Distributors or Non-EU Traders

Non-EU Distributor/Non-EU Trader	• is not allowed to register at all

Although a Non-EU Distributor or a Non-EU Trader will never have the duties of a registrant and REACH is not applicable outside of the EU, there may be some communication obligations within a certain supply chain that are recommended to be fulfilled also by globally acting Non-EU Distributors or Non-EU Traders to support their customers.

Main task	• Pass on information up and down the Supply Chain. Relevant only when there are EU customers or EU suppliers within a certain Supply Chain.

Duties and tasks	Hints
• Clarify whether a customer can benefit from being a Downstream User, because the Non-EU manufacturer takes care via his Only Representative or whether the customer itself acts as an importer and therefore has its own registration obligations	→ If there is no information to the contrary it must be assumed that the importers have registration obligations
• If requested by the Non-EU Manufacturer taking care of the registration obligations via an Only Representative, pass down request concerning uses of the customers	→ Sample letter in Chapter 5, Figure 5.1
• If an upstream supplier requests an Article 18(4) confirmation the Distributor or Trader should request his customers to provide him with such a confirmation	→ Sample letter see Chapter 5, Figure 5.3
• Provide SDS/eSDS either including registration number or without registration number (depending on the role of the customer) and if so updates. SDS/eSDS may include contact details and address of OR to ensure that customers are informed about the appointment of the OR as requested by Article 8 (3) [22] • Alternatively: Duty to communicate information down the supply chain for substances on their own or in preparations for which a safety data sheet is not required	→ REACH Article 31 [5] → Article 8 (3) [13] → REACH Article 32 [6]

Figure 14.15 Communication from Non-EU Distributors or Non-EU Traders to their EU customers.

14.4.1
Communication with EU Customers

After having made the decision in cooperation with the upstream suppliers whether they take care of the registration obligations and are also willing to let the Non-EU Distributors customers benefit from being Downstream users, it may be clarified which role under REACH the customers of a Non-EU Trader or Non-EU Distributor will have. Depending on that outcome the communication from Non-EU Distributor or Non-EU trader to their EU customers will have to be determined.

In Figure 14.15 there are listed duties and tasks that a Non-EU Trader or Non-EU Distributor may have down the supply chain under the assumption that the upstream supplier(s) takes care of the registration obligations.

14.4.2
Communication with Non-EU Manufacturers

Duties and tasks that a Non-EU Distributor or Non-EU Trader may have up the supply chain are listed in Figure 14.16. The role of the Non-EU Distributor or Non-EU Trader will often be similar to that of a Downstream Supplier within the EU. If the Non-EU Distributor or Non-EU Trader supplies a certain substance from a European source it is clear that he may act as a Downstream Supplier having in general the role of a Downstream User under REACH. If the upstream Supplier is a Non-EU source first it needs to be clarified whether a Non-EU manufacturer is willing to take over the registration obligations. A sample letter to ask the Non-EU supplier for clarification can be found in Figure 14.17.

Duties and tasks	Hints
• Request whether Non-EU Supplier (Non-EU manufacturer) will take care of the registration obligations via an Only Representative by also covering volumes that are bought by Non-EU Distributor to be forwarded to EU customers later on	→Sample letter in Figure 14.17
• Pass on information provided by customer concerning their uses to have them made identified uses to be covered in the registration of the Non-EU manufacturer´s Only Representative	→Article 37 (2) [16]
• Pass on information from customers concerning hazards of the substance to the Non-EU manufacturer	→ Article 34 of the REACH regulation [1]
• Provide Non-EU manufacturer with Article 18(4) confirmation on request, if a substance purchased is forwarded as a whole or in parts to EU customers that provided the Non-EU Trader or Distributor with such a document	→Adapt Sample letter from Chapter 5, Figure 5.3 by stating that your company has the status of a Downstream Supplier and therefore gives the confirmation on behalf of your EU customers; Article 18(4) confirmations from your EU customers shall be stored within your company, but must not be passed on to the upstream supplier in respect of the competition law

Figure 14.16 Communication from Non-EU Distributor or Non-EU Trader to an upstream Non-EU manufacturer.

> Dear Supplier,
>
> We would like to ask you whether it would be convenient for your company to cover in your REACH registration also volumes that are sold to our company and may later on be forwarded to our EU customers. We would highly appreciate to have your support on this. Furthermore, we are willing to pass on information to your Only Representative concerning volumes that we forward to EU customers.
>
> Kind regards,
> [company name]

Figure 14.17 Sample letter that may be used to ask a Non-EU manufacturer acting as upstream supplier whether he is willing to also cover volumes forwarded by a Non-EU Trader or Non-EU Distributor to his EU customers.

Duties and tasks	Hints
• Provide OR with volumes that are forwarded to EU customers and provide OR with the identity of these importers to make them Downstream Users	→Article 8 (2) [12]
• If so inform OR about further Non-EU Downstream suppliers that may forward parts of the substance to European customers, if they shall benefit from not generating registration obligations at their EU customers	→Article 8 (2) [12]
• Ask OR to provide EU customers with REACH Certificate	→Sample documents in Section 14.6, Figure 14.25 and Figure 14.26

Figure 14.18 Communication from Non-EU distributor or Non-EU trader to an Only Representative acting on behalf of an upstream Supplier (Non-EU manufacturer).

14.4.3
Communication with Only Representative Acting on behalf of a Non-EU Manufacturer (Supplier)

If a Non-EU manufacturer (upstream supplier) is willing to take over the registration obligations also for volumes supplied to Non-EU customers that may forward the substance to their EU customers, it may be necessary for a Non-EU Distributor or a Non-EU Trader to cooperate with the Only Representative of the Non-EU manufacturer to ensure that his EU customers will not have to take the obligations of an importer under REACH. Figure 14.18 shows what type of duties and tasks may occur.

14.5
Communication to be Done by a Downstream User or a Downstream Supplier

Downstream User	• natural or legal person established within the EU who has neither the role Manufacturer nor the role importer [14]; • re-importer is regarded as a Downstream User [14].
Downstream Supplier	• for example, Distributor, who is not a Downstream User [14], because he does not really use a substance in the meaning of REACH, but forwards the substance to customers down the supply chain.

14.5.1
Communication with Suppliers

The communication to be done by Downstream Users or Downstream Suppliers is very often the same or at least similar.

Figure 14.19 shows tasks that may be fulfilled by a Downstream user and also a Downstream Supplier. In the case of communication to be done with upstream suppliers the tasks will be the same for Downstream User and Downstream Supplier, but for Article 18(4) confirmations the Downstream Supplier will need to adapt the confirmation that he provides to his supplier in so far as he does not use the substance itself, but passes on the information received from his customers.

In accordance with Article 39 (1) [17] Downstream users have to comply with the requirements of Article 37 [18] at the latest 12 months after receiving a registration number communicated to them by their suppliers in a SDS.

The content of the communication from a Downstream User to his supplier will be independent of the fact whether this Supplier is established within the EU or outside of the EEA. However, there may be situations that demand that a Downstream User also communicates with the Only Representative of a Non-EU manufacturer.

Duties and tasks	Hints
• may provide information to assist in the preparation of a registration	➔ Article 37 (1) [15]
• may make use known in writing to the manufacturer	➔ Article 37 (2) [16]
• Provide EU suppliers on request with Article 18(4) [4] confirmation	➔ Sample document in Chapter 5, Figure 5.3.1 which figure is meant here (needs to be adapted for a Downstream Supplier)

Figure 14.19 Communication of a downstream user to his upstream supplier.

14.5.2
Communication with Only Representative of Non-EU Manufacturer

As a Downstream User buying substances or preparations from Non-EU sources shall have a proof of being not an importer under REACH, but really a Downstream User, it seems necessary that the Downstream User is aware of the fact who acts as an Only Representative for the Non-EU manufacturers he purchases from. After having this information the Downstream User should ask the Only Representative for a confirmation and demand for example, a REACH Certificate as shown in Section 14.6 (Figure 14.25 and Figure 14.26).

In the case of indirect imports it may be required that a Downstream User or Downstream supplier contacts the Only Representative also to provide him with information requested under Article 8(2) of the REACH regulation.

An overview on the most important situations when a Downstream User will contact an Only Representative is given in Figure 14.20.

Duties and tasks	Hints
• Contact OR and ask for proof that he takes over registration obligations on behalf of the Non-EU manufacturer and therefore the importer has the role of a Downstream User	→Sample documents in Section 14.6, Figure 14.25 and Figure 14.26
• Provide OR with volumes that are forwarded to EU customers and provide OR with the identity of this importers to make them Downstream Users	→Article 8 (2) [12]
• If so inform OR about further Downstream suppliers that may forward parts of the substance to European customers, if they shall benefit from not generating registration obligations at their importers	→Article 8 (2) [12]

Figure 14.20 Communication from Downstream User, Only Representative.

14.5.3
Communication with Customers (Downstream Users)

The Communication from Downstream Suppliers to their customers is in principle the same as the communication to be done by an EU manufacturer to his customers, therefore see Sections 14.2.1 and 14.2.2 to define tasks to be done in this field.

14.5.4
Communication from Downstream User with Workers

Obligations for granting access to certain information to workers is in general the same as for an EU manufacturer, therefore see Section 14.2.3.

14.5.5
Communication from Downstream User to Authorities

A Downstream User under REACH may have to communicate with different authorities. Besides from on-site inspections or any request in which national authorities will have to be provided with certain information, there may also occur situations in which a supplier will have to communicate directly to ECHA although he is a Downstream User for a certain substance.

In Figure 14.21 there is given information when a Downstream User will be required to prepare their own CSR and has to report to ECHA.

As defined in Article 39 (2) [23] Downstream users shall be required to comply with the requirements of Article 38 [22] at the latest six months after receiving a registration number communicated to them by their suppliers in a safety data sheet [23].

Duties and tasks	Hints
• If an identified use cannot be considered by the registrant, the Downstream User shall prepare his own CSR	→Article 37 (3) [19] →Article 37 (4) [20], Exemptions Article 37 (4) (a) to (f) [21]
• Obligation to report to ECHA	→Article 38 [22]
• If a substance that is subject of authorization is used by a Downstream User he has the obligation to notify ECHA within three months of the first supply of the substance	→Article 66 [10]

Figure 14.21 Communication of a Downstream User to ECHA.

14.6
Communication to be Done by an Only Representative

Only Representative	• natural or legal person established within the EU who after appointment by a Non-EU manufacturer fulfils the registration obligations of an importer.

Duties and tasks	Hints
• Ask for Appointment letter	→ Article 8 (1) [11]
• Ask for SVT, volumes brought to EU, customers sold to and information on the latest update of the SDS for a certain substance	→ Article 8 (2) [12]
• Ask for all information that is required to prepare the registration	

Figure 14.22 Communication from Only Representative to the Non-EU manufacturer who appointed him.

14.6.1
Communication with Non-EU Manufacturer

As there will be ongoing negotiations and discussions between a Non-EU manufacturer and the Only Representative acting on behalf of this Non-EU manufacturer, in Figure 14.22 only the most important subjects of this cooperation are considered in which an Only Representative will communicate to his customer.

Within a company group having different legal entities established within and also outside of the EU that have a good functioning IT-system that may be accessed by all sites, the Only Representative may have direct access to all necessary information via the IT-system, whereas an independently working consultant company acting as an Only Representative may have to ask for much more documents and information.

14.6.2
Communication with ECHA and National Authorities

Besides the fact that an Only Representative in the role of the registrant submits to ECHA all types of dossiers that are required to fulfill the obligations under REACH and consequently has to take care of all correspondence with ECHA, there are further obligations that may require communication. These duties can be found in Figure 14.23 in connection with the relevant Articles of the REACH regulation.

14.6.3
Communication with Customers of the Non-EU Manufacturer

Customers of a Non-EU manufacturer who takes care of the registration obligations via an Only Representative, may be interested in having proof of that. Therefore, they may request from the Only Representative a Certificate concerning the REACH status of a product purchased from the Non-EU Manufacturer. The Only Representative will provide direct customers of the Non-EU manufacturer with a pre-registration Certificate or a registration certificate that the customers can show

Duties and tasks	Hints

• Keep available and up-to-date information on
   • quantities imported;
   • customers sold to;
   • information on the supply of the latest update of the safety data sheet.
	→Article 8 (2) [12] in connection with Article 36 of the REACH regulation [1]
	
	→Article 31 [5]

• If required that a chemical safety assessment is carried out, Only Representative as the registrant shall ensure that the information in the Safety Data Sheet is consistent with the information in this assessment.	→Article 31 (2) and Article 14 of the REACH regulation [1]
→Upon request this information has to be made available to any competent authority of the Member State in which the Only Representative is established or to ECHA	→Article 36 of the REACH regulation [1]

Figure 14.23 Duties and communication obligations of an Only Representative towards authorities.

to national authorities during on-site inspections to demonstrate that they are acting in legal compliance. The Only Representative will provide only customers of the Non-EU Manufacturer on whose behalf he is acting with the corresponding Certificate. This means the Only Representative will either ask the Non-EU manufacturer for a confirmation that a party requesting a certificate is an active customer or the Non-EU Manufacturer provides the Only Representative on a regular basis with a list of his customers. Within a company group this task may cause less expenditure as the Only Representative may have direct access to the company's data base.

However, the situation would be much more complicated when the Non-EU Manufacturer also allows that his Non-EU customers forward a substance to their EU customers by using his registration. Then, the situation may occur that the Only Representative would have to also provide parties that are not known as a direct customer of the Non-EU manufacturer with such a Certificate. It would be difficult to exclude that anybody could abuse this complex situation.

Figure 14.24 shows some occasions when an Only Representative may communicate to the customers of a Non-EU manufacturer who appointed him as an Only Representative. Figure 14.25 shows an example for a pre-registration certificate and in Figure 14.26 there is given an example for a Registration Certificate that may be sent from an Only Representative to the EU-customers of a Non-EU manufacturer on request.

Duties and tasks	Hints
• Inform customers about Appointment as an Only Representative	→In accordance with Article 8 (3) [13] this task shall be done by the Non-EU manufacturer (see sample letter in Chapter 14, Figure 14.10), but this will only work in the case of direct imports. →If indirect imports shall also be covered, it may be necessary that the Only Representative sends either this letter on request of the Non-EU exporter within the supply chain or on request by the EU importer within this supply chain
• Provide customers with preregistration Certificate	→Sample letter in Figure 14.25
• Provide customers with information concerning REACH status of a certain substance on request	
• Provide customers with registration Certificate	→Sample letter in Figure 14.26

Figure 14.24 Communication from Only Representative to customers of the Non-EU Manufacturer that he represents.

REACH PRE-REGISTRATION CERTIFICATE

In compliance with Article 8 of the Regulation (EC) No 1907/2006 of the European Parliament and of the Council of 18 December 2006, concerning the Registration, Evaluation, Authorisation and Restriction of Chemicals (REACH), we *[Only Representative name and address]* were appointed by the Non-EU Manufacturer *[Name and address of Non-EU Manufacturer]* to act as its Only Representative.

[Only Representative name and address] can confirm that we have completed the pre-registration for the following substance(s):

Substance	CAS	EINECS	Tonnage band	Registration deadline	Remarks

If you purchase these substance(s) from *[Name and address of Non-EU Manufacturer]* you will benefit from being a Downstream User without having a registration obligation.

Date: _____   Signature:_____
[Only Representative name]

Figure 14.25 Pre-registration Certificate.

14.7 Examples and Exercises

REACH REGISTRATION CERTIFICATE

In compliance with Article 8 of the Regulation (EC) No 1907/2006 of the European Parliament and of the Council of 18 December 2006, concerning the Registration, Evaluation, Authorisation and Restriction of Chemicals (REACH), we *[Only Representative name and address]* were appointed by the Non-EU Manufacturer *[Name and address of Non-EU Manufacturer]* to act as its Only Representative.

[Only Representative name and address] can confirm that we have completed the registration for the following substance(s):

Substance	CAS	EINECS	Type of registration	Registration number	Remarks
			TII	01-...	Art. 18(4) confirmation required

If you purchase these substance(s) from *[Name and address of Non-EU Manufacturer]* you will benefit from being a Downstream User without having a registration obligation.

Date: _____   Signature: _____
[Only Representative name]

Figure 14.26 Registration Certificate.

14.7
Examples and Exercises

1. When does a Downstream Supplier have the obligation to prepare a Downstream User report?

2. How long must information that is required by a manufacturer, importer, downstream user or distributor to carry out his duties under the REACH regulation be stored?

3. Who has to announce to the customers that a Non-EU manufacturer has appointed an Only Representative?

   ☐ Only Representative ☐ Non-EU manufacturer ☐ ECHA ☐ Nobody

4. Assume that a Non-EU manufacturer produces substance A. Parts of the manufactured volume are sold directly to EU customers, whereas another part is sold to Non-EU Traders that may sell the substance also to EU customers by intending to do that by using the registration number that the Only Representative of the Non-EU manufacturer received by ECHA.

   Which problems may be connected to indirect imports?

5. What sort of information have authorisation holders and also Downstream Users to mention on the label before they place on the market a certain substance or preparation?

6. What communication obligation has a Downstream User against ECHA if he uses a substance that is subject of authorisation and which was an authorisation granted for?

7. What sort of information has to be forwarded from employers to workers and their representatives?

References

1 Regulation (EC) No 1907/2006 of the European Parliament and of the Council of 18 December 2006, concerning the Registration, Evaluation, Authorisation and Restriction of Chemicals (REACH), establishing a European Chemicals Agency, amending Directive 1999/45/EC and repealing Council Regulation (EEC) No 793/93 and Commission Regulation (EC) No 1488/94 as well as Council Directive 76/769/EEC and Commission Directives 91/155/EEC, 93/67/EEC, 93/105/EC and 2000/21/EC.
2 See Article 36 of the REACH regulation [1].
3 See Article 36 (1) of the REACH regulation [1].
4 See Article 18 (4) of the REACH regulation [1].
5 See Article 31 of the REACH regulation [1].
6 See Article 32 of the REACH regulation [1].
7 See Article 33 of the REACH regulation [1].
8 See Article 35 of the REACH regulation [1].
9 See Article 65 of the REACH regulation [1].
10 See Article 66 of the REACH regulation [1].
11 See Article 8 (1) of the REACH regulation [1].
12 See Article 8 (2) of the REACH regulation [1].
13 See Article 8 (3) of the REACH regulation [1].
14 See Article 3 (13) of the REACH regulation [1].
15 See Article 37 (1) of the REACH regulation [1].
16 See Article 37 (2) of the REACH regulation [1].
17 See Article 39 (1) of the REACH regulation [1].
18 See Article 37 of the REACH regulation [1].
19 See Article 37 (3) of the REACH regulation [1].
20 See Article 37 (4) of the REACH regulation [1].
21 See Article 37 (4) (a) to (f) of the REACH regulation [1].
22 See Article 38 of the REACH regulation [1].
23 See Article 39 (2) of the REACH regulation [1].

Appendix – Answers and Solutions Concerning the Sections Examples and Exercises within this Book

Chapter 2

1. Only company A has registration obligations as a European manufacturer. Company B manufactures outside EU and therefore has no obligation to register substance S.

2.
   a) Only company A as a manufacturer within the EU has registration obligations.

   b) Company A (EU manufacturer) and customer C (importer for the amount purchased from company B) – alternatively company B could register via an OR in the EU, then Company C is a Downstream User also for the amounts of substance S purchased from company B.

   c) Only company A as a European manufacturer that manufactures more than 1 t/a of substance S. Concerning amounts of substance S that a customer located in the European Union purchases from Manufacturer B in general the European customer in the role of the European importer has to take care of the registration obligations, but if the total amount of S that is imported per calendar year is below 1 t/a for each importer, nobody will be obliged to register substance S.

3. There are several options to cope with the situation and to achieve legal compliance. Here we give two options, but maybe you will find some other solutions for your company in similar situations.

   Option 1: Company A purchases 2 t of substance S just in time in January 2012. Company A then has the role of the importer under REACH and imports more than 1 t/a. Company A has to register the substance S before the import. It is highly probable that company A is not in the position to register in due time.

   Option 2: Company A purchases 999 kg of S in December 2011 and again 999 kg of S in January 2012. When company A does not purchase further amounts of

S that are not registered yet, company A does not need to register at all, because the import had been less than 1 t per calendar year.

Chapter 3

1.

Product	Substance, preparation or article	Reasons
benzene	substance	In accordance with the definition in REACH Article 3 (1) [1]
bottle	article	In accordance with REACH Article 3 (3) [2]
chair	article	In accordance with REACH Article 3 (3) [2]
chlorine	substance	In accordance with the definition in REACH Article 3 (1) [1]
Dodecanediocic Acid	substance	In accordance with the definition in REACH Article 3 (1) [1]
Hydrogen chloride	substance	In accordance with the definition in REACH Article 3 (1) [1]
Hydrochloric acid	preparation	Hydrogen chloride diluted in water, both substances can in principle be separated again
Iodine	substance	In accordance with the definition in REACH Article 3 (1) [1]
methanol	substance	In accordance with the definition in REACH Article 3 (1) [1]
Mixture of methanol and ethanol	preparation	Both substances can be separated by distillation without affecting the stability of the two substances methanol and ethanol, see also [3]
Oleum	preparation	SO_3 diluted in sulfuric acid, see also [3]
Plastic knife	article	In accordance with REACH Article 3 (3) [2]
Sodium chloride	substance	In accordance with the definition in REACH Article 3 (1) [1]
Solution of iodine in benzene	preparation	Solvent may be separated, see also [3]
Sulfuric acid (100%)	substance	In accordance with the definition in REACH Article 3 (1) [1]
Sulfur trioxide	substance	In accordance with the definition in REACH Article 3 (1) [1]

Appendix – Answers and Solutions Concerning the Sections Examples and Exercises within this Book

Product	Substance, preparation or article	Reasons
tetrahydrofurane	substance	In accordance with the definition in REACH Article 3 (1) [1]
Vacuum cleaner	article	In accordance with REACH Article 3 (3) [2]
Xanthydrol (10% in Methanol)	preparation	Solvent may be separated without affecting the stability, see also [3]

2.

a) → mono-constituent substance.

b) → mono-constituent substance, as water does not have to be considered. In the water-free product the amount of sulfuric acid is above 80% and therefore it is a mono-constituent substance.

c) Quality Excellent: As it contains a minimum 80% of Substance S it can be regarded as a mono-constituent substance.

Quality Normal: As it contains less than 80% of Substance S it must be regarded as a multi-constituent substance.

Concerning the registration it is in any case possible to register both qualities separately. As this will lead to higher costs (two dossiers, two times fees for registration, etc,) a company would rather submit only one registration dossier, especially when the uses and exposure scenarios for both qualities are similar, physico-chemical properties are similar and maybe classification and labeling is identical. Maybe the most comfortable way to cope with the situation is to register Substance S as a mono-constituent substance and state that there is also another quality available. However, the Quality Normal deviates from the 80% rule. Alternatively, a registration as reaction mass can be considered. In this case, the composition can be given in such figures that the concentration range for both qualities is covered by stating a wider range for each constituent than for the single quality. Each constituent in amounts of 10% and above will be listed as a constituent, those below 10% have to be defined as impurities within the Substance Data Set/registration dossier.

Whenever you are in doubt concerning a similar situation for a substance manufactured by your company, please do not hesitate to ask ECHA for their support.

d) → mono-constituent substance, as water must not be considered – therefore in the water-free part of the product the hydrogen chloride content will be above 80%.

3.

Substance	CAS	Phase-in substance	NLP	Non-phase-in substance
2-methylisocrotonic acid (Angelic acid)	565-63-9	EC 209-284-2	–	–
Angelic acid methyl ester	5953-76-4	EC 227-718-9	–	–
Angelic Anhydride	94487-74-8	–	–	x
2-Bromopropionyl bromide	563-76-8	EC 209-261-7	–	–
Benzyl alcohol	100-51-6	EC 202-859-9	–	–
D-(+)-Turanose	547-25-1	EC 208-918-5	–	–
Biotin	22879-79-4	–	–	x
n-Valeryl bromide	1889-26-5	–	–	x
n-Valeryl chloride	638-29-9	EC 211-330-1	–	–
Acetic acid	64-19-7	EC 200-580-7	–	–
Sulfuric acid	7664-93-9	EC 231-639-5		
Butene, dimers	9021-92-5	–	EC 500-025-1	–
Hexamethylene diisocyanate, oligomers	28182-81-2	–	EC 500-060-2	–
4-amino-6-[(1,2-dihydro-2-imino-1,6-dimethyl-4-pyrimidinyl) amino]-1,2-dimethylquinolinium hydroxide	93919-18-7	EC 300-002-4	–	–
Phenol, propoxylated	28212-40-0	–	EC 500-61-8	–
2,2'-[(4-[2-hydroxyethyl) amino]-2-nitrophenyl]imino] bisethanol	93919-22-3	EC 300-007-1	–	–
Aluminum isopropoxide	555-31-7	EC 209-090-8	–	–
Oleic acid, ethoxylated	9004-96-0	–	EC 500-015-7	–
Vanillin	121-33-5	EC 204-465-2	–	–
Xanthydrol	90-46-0	EC 201-996-1	–	–

The reason for decision is in any case the EC number connected to the stated CAS number. EC 2xx-xxx-x and 3xx-xxx-x show that a certain substance is listed in EINECS and therefore it is a phase-in substance. EC 5xx-xxx-x indicates NLPs – they can be treated as phase-in substances under REACH. No EC number or EC numbers in the format 4xx-xxx-x, 6xx-xxx-x, 7xx-xxx-x 9xx-xxx-x indicate that a substance is a non-phase-in substance.

Appendix – Answers and Solutions Concerning the Sections Examples and Exercises within this Book | 259

4. The substance Angelic Anhydride is a non-phase-in substance. Pre-registration of such a substance was only possible in accordance with Article 3 (20) (b). My company had to be within the EU and must have produced or imported Angelic Anhydride at least once in the 15 years before REACH entered into force, but the substance was not placed on the market by my company (e.g., because it was used as an intermediate in the synthesis of another substance). Therefore, in accordance with REACH Article 3 (20) (b) my company could have claimed phase-in status for this substance (although it is not listed in EINECS) before the deadline for pre-registration on 30 November 2008. No late pre-registration can be done for that substance, as either my company had to do the pre-registration because of Article 3 (20) (b) in advance of the deadline 30 November 2008 or the substance has to be considered as a non-phase-in substance and treated as such.

5. In accordance with REACH Annex V (9) nitrogen is exempted from registration obligations as hazards and risks for this substance are already well known. Therefore, it was not necessary/possible to pre-register nitrogen.

6.
   a) Yes. Hydrogen, CAS [1333-74-0], EC [215-605-7] is according to REACH Annex V (9) exempted from the obligation to register, because hazards and risks for this substance are already well known.

   b) Yes, it is different. In general for by-products that are isolated and afterwards shall be marketed a company has registration obligations.

   c) If your company is not located within the EU there are no registration obligations concerning REACH.

   It is assumed that your company is located within the EU:

   When the substance is isolated and afterwards used for the synthesis of another substance it may be an on-site isolated intermediate for your company and there are registration obligations at least for an on-site isolated intermediate (depending on whether your company is able to handle the substance in accordance with REACH Art. 17 (3)).

   When the substance shall not be used in further processing but is burned, it can be defined as waste and therefore it is exempted from REACH even if it has to be transported to another site before it can be burned.

Chapter 4

1. The company in Germany has to register the amounts of substance A that are manufactured at their own site, which means 50 t/a. The French manufacturer has also to register the amount of substance A that is manufactured at

his site per calendar year. When the French manufacturer manufactures more than the 10 t/a that are delivered to the German company, then they have to consider the whole amount of substance A that is produced at their site per year.

2. The German company has to register in any case the amounts of substance A that are manufactured at their own site. The amounts of substance A that are purchased from the Chinese manufacturer can be covered in the same registration dossier, if the German company registers not only in the role of the Manufacturer but also in the role of an importer under REACH. Then, the total amount manufactured and imported is relevant for the tonnage band. As the tonnage band will be 10 t/a to 100 t/a in both cases (Manufacturer only, Manufacturer and Importer) data requirements will be the same for both cases.

If the Chinese Manufacturer appointed an Only Representative within the EU who will register on behalf of the Chinese Manufacturer, the Only Representative will consider in his registration dossier the amount of substance A that is delivered to EU customers from the Chinese Manufacturer. In this case, the German company purchasing 10 t/a of substance A from this Chinese Manufacturer can benefit from being a Downstream User for the amounts of substance A that are purchased from the Chinese Manufacturer. The German company then has to consider in their own registration dossier only the amounts of substance A that are manufactured at their own site per calendar year.

3. If the substance shall be manufactured in amounts of 1 t/a and above there will occur registration obligations. If the substance is manufactured in amounts of less than 100 t/a and the substance is not a CMR substance the chemical company may do a late pre-registration and afterwards benefit from an extended registration deadline until 31 May 2018. In all other cases, the chemical company has to register immediately. "To register immediately" means in any case to submit at first an Inquiry dossier to ECHA and afterwards submit a registration dossier either on your own or as a Member of Joint submission. The rules then are equal to those that are valid for non-phase-in substances.

4. As the raw material is purchased from a Non-EU manufacturer (company located in the United States) there will occur registration obligations concerning the import of substance A to a European site. Either EXTRACLEVER Germany or EXTRACLEVER France could act as the importer under REACH. Alternatively, the Manufacturer from the United States could appoint an Only Representative within the EU. Then, EXTRACLEVER sites located within the EU would have the role of a Downstream User without registration obligations concerning the import of substance A.

If a Non-EU site of EXTRACLEVER will import substance A from the United States, nobody will have registration obligations, because the substance will be used completely for the transformation into another substance and therefore substance A will not be handled within the EU.

If the intermediates B and C will not be isolated, there will be no registration obligations at all, even if the manufacture of D takes place at a European site. If the intermediates B and/or C are isolated at a European site, they have to be registered at least as an on-site isolated intermediate. Non-European manufacturers do not have to register such on-site isolated intermediates.

Concerning the product D every EU manufacturer will have the obligation to register this substance, whereas the Italian customer can benefit from being a Downstream User. If EXTRACLEVER manufactures substance D outside of the EU (in China or in Switzerland) there are two possibilities to achieve REACH compliance concerning the registration obligations for Substance D. Either the Italian customer takes care in the role of the importer or EXTRACLEVER (China or Switzerland) appoints an Only Representative within the EU. If EXTRACLEVER takes care of the registration obligations they may bind the customer and they may realize it soon if there would be competitors also having registered substance D or willing to register this substance. On the other hand, if the Italian customer takes care of the registration obligations he will be in the position to buy not only from EXTRACLEVER (China or Switzerland) but also from any other trader or manufacturer outside of the EU.

As Switzerland is at the shortest distance to the Italian customer in addition to the advantageous situation concerning REACH tasks it may be a good idea to manufacture at the Swiss site of EXTRACLEVER.

	China	Germany	France	Switzerland
Site within or outside of the EU	Non-EU	EU	EU	Non-EU
Registration obligations concerning raw material A	No	Yes	Yes	No
Registration obligation concerning non-isolated intermediates B and C	No	No	No	No

(Continued)

	China	Germany	France	Switzerland
Registraion obligations concerning isolated intermediates B and C	No	Yes–maybe competitors can trace back to manufacturing process after ECHA published certain information on their website	Yes–maybe competitors can trace back to manufacturing process after ECHA published certain information on their website	No
Registration obligations concerning product D	No (Italian customer has the obligation to register as an importer)	yes	yes	No (Italian customer has the obligation to register as an importer)
Distance for transport of product D to the customer	Great distance	acceptable	acceptable	shortest distance

Chapter 5

1. As the substance is used as a catalyst it is clear that a standard registration is required even when the substance is manufactured at your own site. If the substance is purchased from a Non-EU supplier it will be the same, but your company will register in the role of an importer. If the substance is purchased from an EU supplier or from a Non-EU supplier who takes care of the registration obligations via an Only Representative there may be no obligation for your company to register if the registrant covers your use in his registration dossier.

   If a supplier registers the mentioned substance, but does not cover the use of your company there may occur the situation that your company has to prepare in the role of the Downstream User your own Chemical Safety report in accordance with REACH Article 37 (4) [4].

2. In any case, it will be possible to do a standard registration as all uses can be covered.

   For an intermediate it should be checked, whether it is possible to submit a registration dossier for a transported isolated intermediate. This inexpensive type of registration requires that the substance is completely used for the synthesis of (an)other substance(s) and the substance is handled under strictly controlled conditions as defined in Article 18 (4) of the REACH regulation for the whole lifecycle. The registrant will have to ask all his customers

(Downstream Users) for a written confirmation ("Article 18 (4) confirmation") that they handle the substance as demanded in REACH Article 18 (4).

3. For a Non-EU manufacturer, for example a Swiss Company there will occur no registration obligation at all in such a case.

   For a EU manufacturer it has to be checked whether the manufactured substance is a non-isolated intermediate or an on-site isolated intermediate. Non-isolated intermediates must not be registered, whereas an on-site isolated intermediate in any case has to be registered by a EU manufacturer. If the on-site isolated intermediate is handled under strictly controlled conditions during its whole lifecycle as mentioned in Article 17 (3) of the REACH regulation, the EU manufacturer may do a registration as an on-site isolated intermediate and benefit at least from reduced data requirements compared to a standard registration. If the substance is used for the synthesis of another substance, but is not handled under strictly controlled conditions in accordance with REACH Article 17 (3) the manufacturer will have to prepare a standard registration.

4.

Type of registration	May Only Representative register on behalf of the Non-EU manufacturer?	Justification
Standard registration (full registration/ registration as a substance)	yes	In the case of import of the substance into the EU the OR may take care of the registration obligation instead of the importer. EU customers then will be Downstream Users without their own registration obligations.
Registration as a transported isolated intermediate (TII)	yes	In the case of import of the substance into the EU the OR may take care of the registration obligation instead of the importer. EU customers then will be Downstream Users without their own registration obligations.
Registration as an on-site isolated intermediate (OII)	no	There is no registration obligation as the substance is neither manufactured within the EU nor imported into the EU. The area of application of the REACH regulation is Europe (EEA).

5. As an Only Representative can only act on behalf of Non-EU manufacturers but not on behalf of any Non-EU trader this will not be possible at all.

6. Usually at universities substances are manufactured for the purpose of Scientific Research and Development in accordance with REACH Article 3 (23) and amounts are below 1 t/a. Therefore, no registration obligation will occur.

7. There are several options to fulfill the registration obligations:

   Option 1: Standard registration of the whole amount that is manufactured. That means registration in the tonnage band 100 to 1000 t/a and data requirements according to REACH Annex VII, VIII and IX (concerning data requirements see also Chapter 6 of this book).

   Option 2: Standard registration for the tonnage band 10 to 100 t/a by fulfilling data requirements as described in REACH Annexes VII and VIII <u>AND</u> additionally a registration for a transported isolated intermediate in the tonnage band 1 to 10 t/a (TII < 1000 t/a).

   →Option 2 will be cheaper (lower fees with ECHA and lower expenses for studies/tests) but Substance Volume Tracking (SVT) must distinguish between standard registration and transported isolated intermediate. Furthermore, for the amount sold as a transported isolated intermediate an "Article 18 (4) Confirmation" from the customer(s) is required.

8. As substance A is imported into the EU and afterwards used for the synthesis of another substance it may be registered as a transported isolated intermediate if the importer will handle the substance in accordance with the strictly controlled conditions as described in Article 18 (4) of the REACH regulation.

   As substance B is transported to another site and there used for the synthesis of another substance it may be correct to register this substance also as a transported isolated intermediate. The manufacturer of substance B has to manufacture B under strictly controlled conditions as described in REACH Article 18 (4) and he has to ask the company that uses B for the manufacture of C for their confirmation that they use substance B accordingly.

   When substance C is not isolated, there will be no registration obligations as non-isolated intermediates are exempted from registration obligations. If the European company that manufactures substance C, will isolate substance C previous to further transformation into substance D there will occur registration obligations. Substance C has to be registered at least as an on-site isolated intermediate under the condition that the substance is manufactured and used under strictly controlled conditions, as described in REACH Article 17 (3) for on-site isolated intermediates.

   Substance D has to be registered by the manufacturer as a transported isolated intermediate for that part that is used as an intermediate by the customer when it is manufactured and used under strictly controlled conditions in accordance with Article 18 (4) of the REACH registration and the manufacturer of substance D has received a confirmation from the customer/Downstream user that the substance is handled accordingly.

At least for the part of substance D that is used as a catalyst a standard registration is required. The manufacturer of D may either do a standard registration for the whole amount of D that he manufactures or he may split the tonnages and do separate registrations for a transported isolated intermediate and also a standard registration.

Only for the sake of completeness: As we see at a later stage it would also be possible for the manufacturer of D to register only a transported isolated intermediate and describe the use as a catalyst in the section "uses advised against" in his registration dossier. Then, the customer/Downstream User may have the obligation to prepare their own Chemical safety Report in accordance with REACH Article 37 (4) for the amounts of the substance used outside the conditions described in the registration dossier of his supplier.

Chapter 6

1.

a) Either standard registration (full registration) as all uses of the substance can be covered or registration as a transported isolated intermediate when the substance is used for the synthesis of another substance and handled in accordance with REACH Article 18 (4).

b) It may be a standard registration, because all uses of your customers can be covered and you do not have to ask your customers for a confirmation concerning the handling of the substance. However, it is highly likely that one would rather do a registration as a transported isolated intermediate whenever it is possible, because there are fewer data requirements, no CSR will be needed and costs for preparation of the dossier and also fees from ECHA are lower than in the case of the full registration. Furthermore, for a transported isolated intermediate a company in any case will benefit from the fact that this substance will not be subject of authorization in future (see REACH Article 2 (8) (b)), whereas substances registered with a standard registration could become subject of authorization.

c) For a standard registration data requirements are based on the tonnage band. For the tonnage band 100 to 1000 t/a there are data requirements as described in REACH annexes VII, VIII and IX to be fulfilled and a Chemical Safety Report has to be provided.

If the substance is registered as a transported isolated intermediate in the tonnage band <1000 t/a there are no special data requirements to be fulfilled.

d) For a standard registration your company has to fill in Sections 1.1, 1.2, 1.3, 1.4, 2.1, 2.3, 3.1., 3.2, 3.3, 3.5, Chapters 4 to 7, Chapter 11 and Chapter 13 (CSR).

For a registration as a transported isolated intermediate your company has to fill in at a minimum (will be checked): Sections 1.1, 1.2, 1.3, 2.1, 3.1 (standard phrase), 3.3, 3.5 and Chapter 11. Data concerning Chapters 4 to 7 shall be entered when they are available.

e) A Member of Joint submission has not to submit Section 2.1 and Chapters 4 to 7 (as the Lead Company does submit these Chapters on behalf of all Members of Joint submission). For a standard registration Chapter 11 and the CSR may also be submitted by the Lead Company on behalf of all Members of the Joint submission, but very often Chapter 11 is submitted by each registrant on his own. Concerning the CSR to be attached to Chapter 13 it may be useful that at least parts of it are submitted by each registrant on his own. If a company wants to add further uses or amend the CSR later on there will be more flexibility in submitting parts or the whole CSR separately. The best solution has to be discussed with the Lead Company.

For a transported isolated intermediate Chapter 11 has to be submitted by each company on their own.

f)

	Single submission	Member of Joint submission (Lead Company is included)
Standard registration in the tonnage band 100 to 1000 t/a	12 317 €	9237 €
Transported isolated intermediate	1714 €	1285 €

2. A manufacturer located within the EU (as described in question 1) has to register such a substance at least as an on-site isolated intermediate (substance has to be manufactured and used in accordance with REACH Article 17 (3)), whereas a Non-EU manufacturer would not have to register such a substance at all.

In case the registrant can prove that the substance is manufactured and handled at his site in accordance with REACH Article 17 (3) he has no special data requirements, but even then ECHA expects that available data concerning IUCLID Chapters 4 to 7 are provided within the registration dossier.

Fees with ECHA will be 1714 € (single submission) or 1285 € (Member of Joint submission).

3. Previous to the submission of a registration dossier an Inquiry dossier has to be submitted to ECHA. Inquiry dossiers are checked in depth by ECHA and

especially IUCLID Section 1.4 has to be filled in with due diligence as ECHA expects a set of analytical data and method descriptions as described in REACH Annex VI, Sections 2.3.5 to 2.3.7 and, furthermore, statements in the case one of the foreseen methods seems not to be appropriate. There is no time line for ECHA. The Check of the Inquiry Dossier may last several weeks. In case of rejection a new Inquiry Dossier has to be submitted by consideration of the special demands from ECHA that they sent within the decision letter. Therefore, it may last several months until you receive an Inquiry number and the list number by ECHA that are required previous to the submission of the registration dossier itself.

4. Yes and no. A PPORD notification can be done for amounts of 1 t/a and above when the conditions stated in REACH Article 3 (22) are met. An upper limit does not exist, but amounts that are manufactured/handled must be for the purpose of PPORD. If your pilot plant cannot be run with smaller amounts it may be allowed to handle several hundred tons per year or even more.

5.
   a) No, in both cases fees with ECHA are 1714 € for the single submission – either for a standard registration or a registration as a transported isolated intermediate.

   b) For a standard registration in the tonnage band 100 to 1000 t/a ECHA invoices 12 317 € (single submission), whereas for the registration of a transported isolated intermediate only 1714 € are due to be paid. Therefore, the difference is 10 603 €.

6. Conditions that have to be considered in entering data and information into a IUCLID Substance Data Set are in short as follows:

   - type of submission (PPORD, Inquiry, registration);
   - Lead Company/single submission or Member of Joint submission;
   - Type of registration (standard, TII, OII);
   - tonnage band (see Figure 6.1) (CSR from 10 t/a, for higher tonnage bands more data have to be filled in in Chapters 4 to 7);
   - Role of the registrant itself M and/or I or OR.

Chapter 7

1. The abbreviation "NONS" stands for **no**tified **n**ew **s**ubstances. NONS are substances that were not listed in EINECS and therefore had to be notified with national authorities under Directive 67/548/EEC if they were marketed. Under REACH they can be regarded as registered for the former notifier, but a registration number has to be claimed from ECHA.

2.
- a) Company A had to notify substance S with the national authorities as the substance was marketed. Company B did not have any obligation to notify Substance S as it was not marketed.

- b) Company A highly likely claimed a registration number as substance S can be regarded as registered under REACH for company A. Therefore, ECHA should have assigned a registration number until 01 December 2008. Maybe in reality it lasted longer, but in principle substance S was registered for company A from that date on at least in the smallest tonnage band. If the previously notification was done for a higher tonnage band the registration is also valid for this tonnage band.

  Company B should have made a pre-registration for Substance S until 30 November 2008 as they have registration obligations for on-site isolated intermediates under REACH, but can benefit from the phase-in status of Substance S in accordance with REACH Article 3 (20) (b), because substance S was "manufactured in the Community, or in the countries acceding to the European Union on 1 January 1995 or on 1 May 2004, but not placed on the market by the manufacturer or importer, at least once in the 15 years before the entry into force of this Regulation, provided the manufacturer or importer has documentary evidence of this;" [5].

3. SNIF File/IUCLID File and notification number. The file had to be imported into your IUCLID Account and then had to be amended to meet the actual requirements in order to prepare an update.

4. [x] spontaneous update

   [x] update on request

5. [x] CLP notification – in the case of formerly notified substances ECHA recommended to update until January 3rd, 2011 instead of doing a separate CLP notification

   [x] increase of tonnage band

   [ ] new customers

   [ ] moving of the national authority

   [x] new knowledge of the risks of the substance to human health and/or the environment

   [x] reasons as stated in REACH Article 22 (1) [6]

   [ ] change in terms of delivery

Chapter 8

1.
   1. First, check whether the amount manufactured per calendar year will be 1 t/a or above. If the amount will be less than 1 t/a there will occur no registration obligations.
   2. If the amount is 1 t/a and above in the second step you should check whether there is an entry in EINECS for a certain substance.
      a) If there is an entry, please check whether a late pre-registration is still possible. If so do a late pre-registration in due time and afterwards treat the substance similar to pre-registered phase-in substances. If late pre-registration is no longer allowed, for example, because the substance is a CMR or R50/53 in combination with a tonnage band above 100 t/a is valid or it is less than 12 months until the registration deadline you have to proceed as for non-phase-in substances and submit an inquiry dossier previous to the registration.
      b) If there is no entry in EINECS (and the substance is also not a NLP) you may assume that it is a non-phase-in substance and therefore an inquiry dossier will be needed previous to the registration.
   3. Submit an Inquiry dossier to ECHA for all relevant substances.

2. IUCLID Section 1.4 Analytical information. This Chapter is very important for ECHA as it may be used to prove the composition of a substance that is important to bring together registrants/potential registrants for the same substance. The Substance Sameness Check has to be based on similar analytical methods for several potential registrants. Substance sameness is a criterion for having Data-sharing obligations and also the possibility to do a Joint submission.

3. A non-phase-in substance must be registered previous to manufacture or import of 1 t/a and above, therefore the tonnage threshold will be reached on the first day of the production. Hence, registration should be done previous to the start of the first production campaign.

   All related activities (preparation of inquiry dossier including all necessary data, especially analytical data and methods, defining of type of registration by considering the use of the substance, maybe preparation of exposure scenarios and Chemical Safety Report) shall be done in due time. Therefore, maybe even if you hurry up with all activities you may need 1 or 2 years for the preparation of the registration dossier, depending on the data requirements necessary for the appropriate type of registration. For the above stated example it may be considered to manufacture instead of 10 t/a an amount slightly below the 10 t/a threshold. If your company has to do a full registration for this substance you may save time and money, because you do not have to provide a Chemical Safety Report for amounts manufactured below 10 t/a. In cooperation with your customer(s) you may also check whether it is possible to benefit from reduced data requirements by doing a registration as a transported isolated intermediate. Alternatively, your company may check whether it is possible to do a PPORD notification previous to the preparation of a registration dossier.

Maybe you cannot give an answer when your company exactly has to start activities to have the registration done in due time, but as in real life the question more often will be what to do to be in legal compliance until a special date to meet your customer(s) demands, it may be that you have to check alternative routes to the originally intended procedure.

4. It is difficult to offer a single solution that will solve all your problems at any time, but as the situation described in the above example may occur very often especially within companies that are offering services in the field of custom manufacturing, it seems to be necessary to list as many alternative options including their consequences to make a decision as to how to proceed in a certain case.

Sometimes, it will be enough to find a solution that gives you some more time until a substance has to be registered or maybe another type of registration will help to reduce data requirements. One could even consider registering first in a smaller tonnage band, start with the production campaign and afterwards do an update because of tonnage band increase. There is no special date until that such an update has to be done, but REACH demands which companies act in such a case without undue delay.

Another option could be to have a PPORD notification for a transition period. Please do not forget that certain conditions have to be fulfilled to benefit from a certain time from this alternative.

If your company also runs sites outside of Europe it may be possible to manufacture a certain substance outside of Europe and sell it to Non-European manufacturers without having any obligation to register under REACH. If the substance shall be delivered to European customers at a later stage your company has the chance to register this substance in the role of the importer (or as an OR for the Non-EU entity), but there may be fewer data requirements if only parts of the manufactured product are delivered to Europe or the import of the whole amount can be split over several calendar years. At least your company is allowed to manufacture the substance without having done the registration and therefore you can win the time that is needed for the production campaign. If the delay in preparing the registration dossier will not be too long this may solve your problem.

The same can be applied in principle if your company asks a Non-European supplier to provide you with a certain substance instead of doing the first campaign at your own European site.

I am sure you will find some more ideas to support your own business.

Chapter 9

1.
   - Data Holders.
   - Third Party Representatives.
   - Potential registrants.

2. It depends on the tonnage band and properties of the phase-in substance. If there are at least twelve months left until the registration deadline it will be possible to do a late pre-registration for this substance. As the latest registration deadline in 2018 will be relevant, there can only be the two smallest tonnage bands from 1 t/a to 10 t/a and 10 t/a to 100 t/a to be considered, if the substance is not a CMR substance.

3. Yes, there is a difference. Although both roles can be notified with ECHA the situation and status will be different. A Candidate Lead Registrant may have notified with ECHA that he intends to become Lead Registrant, but did not do a survey in the (pre-)SIEF yet. After a corresponding survey among the (pre-)SIEF members a company that volunteered to become Lead Registrant/Lead Company may be elected and therefore is allowed to notify ECHA of being the elected Lead Registrant. The elected Lead Registrant will act with the Agreement of the SIEF members.

4. No, even when a registrant can justify to opt-out from parts of a joint submission because Article 11 (3) or Article 19 (2) of the REACH regulation can be applied, it is foreseen that this registrant remains a member of the joint submission.

5. The dossier of the Lead company has to be submitted first and shall at least be accepted for further processing previously to the submission of member dossiers. The Lead company will provide the members of joint submission with submission name and token. Then, the members of joint submission shall enter this information in REACH-IT to determine that they are a member of joint submission. Their registration dossier shall be submitted after having determined their membership in the joint submission. If a member of joint submission would submit their registration dossier before indicating that they are a member of joint submission this will lead at least to increased fees, as ECHA then will invoice the fee for a single submission.

Chapter 10

1. No. As there had been a failure concerning Business rules, the dossier is not accepted for further processing in REACH-IT/at ECHA and therefore you have the obligation to submit a new dossier not an update. It is not possible to submit an update successfully in such a case. Therefore, it is not required to have a submission number of the first trial for the second one.

In general, for Updates the submission number of the last successful submission has to be provided in the dossier header. As the first submission was rejected it was not successfully submitted and therefore in the next trial you have to provide a new dossier without referring to the first one.

2. A reference number is a more general designation than a registration number. A registration number is a reference number that a registrant receives for a certain substance in the case of a successful submission of a registration dossier. If another type of dossier was submitted the submitting party will receive a reference number in accordance with the corresponding dossier type for example, a PPORD notification number for the successful submission of a PPORD notification or a CLP notification number after submission of a CLP notification. After successful submission of an inquiry dossier the submitting party will receive an inquiry number as the corresponding reference number.

3.

Reference number	Information
01-xxxxxxxxxx-28-0000	Registration number, ending on "-0000", therefore either submitted by a Lead Company or as a single submission
01-xxxxxxxxxx-28-0001	Registration number, ending on "-0001", which indicates that there are at least two companies that registered the substance with the BASE NUMBER "xxxxxxxxxx" – the party that received the registration number ending on "-0001" registered as the second company and therefore is a Member of Joint submission
05-xxxxxxxxxx-xx-0000	A certain substance was pre-registered by a certain company in advance of the deadline in 2008 because it was either a phase-in substance (listed in EINECS) or had at least phase-in status for this company
02-xxxxxxxxxx-xx-0000	A company submitted a CLP notification to the CLP inventory. It was required only if the manufacturer of a certain substance did no registration in 2010. Had to be submitted by all EU customers of a Non-EU manufacturer if the Non-EU manufacturer did not register this substance via an Only Representative in 2010. Had to be submitted until the end of 2010 by any importer that imported a certain substance from a Non-EU manufacture if it was intended to bring this substance to market.

4. Inquiry number: 06-xxxxxxxxxx-xx-0000

5. A reference number consists only of digits and hyphens in the format

    2 digits-10 digits-2 digits-4 digits

    <TYPE>-<BASE NUMBER>-<CHECK SUM>-<INDEX NUMBER>

    Whereas a submission number also includes letters in the format

    <2 uppercase letters><6digits>-<2 digits>.

Appendix – Answers and Solutions Concerning the Sections Examples and Exercises within this Book | 273

6.

"Combination of numbers and letters"	Submission number	Reference number	Neither submission number nor reference number – justification
LE900070-68	x	–	
MAX899605-98	–	–	Three uppercase letters instead of two uppercase letters therefore it cannot be a submission number
Pz899643-85	–	–	"z" is not an uppercase letter, therefore it is not a submission number
01-611273564-78-0002	–	–	No reference number, because base number in a reference substance consists of 10 digits, but here there are only 9 digits

7. The reference number/registration number has to be provided in IUCLID "Section 1.3 Identifiers". The submission number of the last successful submission is required for the dossier header enabling REACH-IT to have a reference to the last successfully submitted dossier.

8.
   - Use the fee-calculation plug-in in IUCLID5 to check your dossier.
   - Check the "COMMISSION IMPLEMENTING REGULATION (EU) No 254/2013 of 20 March 2013 amending Regulation (EC) No 340/2008 on the fees and charges payable to the European Chemicals Agency pursuant to Regulation (EC) No 1907/2006 of the European Parliament and of the Council on the Registration, Evaluation, Authorization and Restriction of Chemicals (REACH)".

9. In general, ECHA will respond within a 3-week period. If you are not informed to the contrary within this period your company is allowed to start manufacture (in the case of non-phase-in substances) and there is no need to interrupt the production of a previously pre-registered substance even if you do not receive any information from ECHA until the deadline for registration. The registration deadline for pre-registered substances in the tonnage band 100 t/a up to 1000 t/a (not CMR and not R50/53) is the 31.05.2013. Within the period of the last two months before the registration deadline ECHA will have an extended 3-month period for their response.

That means if the submitting party in the above example submitted the registration dossier successfully by the end of March 2013 it can be ensured that they will have the registration number in due time before the registration deadline.

If the dossier will be submitted on April 1st 2013 there may be a problem even if the dossier is accurately done, as ECHA will have a period of up to three months to check the dossier. The date until the registrant to be may receive any response from ECHA may last until sometime in June 2013 and therefore the registration deadline has passed. In the worst case, a manufacturer has to interrupt the production of this substance beginning on 01 June 2013 until the receipt of the registration number.

Therefore, it is highly recommended to submit your registration dossiers in due time, especially if you are not well experienced in doing dossier submissions, calculate some more time to prepare the dossier with due diligence and also have the chance to correct your dossier in case ECHA demands you to submit further information.

If the dossier was accepted for further processing (passed Enforce Rules Check) the registrant to be will have the chance to correct the dossier once if ECHA demands further information. In any case, the fee has to be paid to ECHA. If ECHA is not satisfied with the information sent on demand the dossier will be rejected and the registrant to be has paid the fee without having received the desired reference number. For the next submission the registrant to be will have to pay the fee again.

Chapter 11

1. None. Change of address can be done in REACH-IT. As this is the "Master file" neither the dossiers for the 22 already registered substances nor the further 56 pre-registrations have to be updated. However, for updates because of any other reason, it is recommended to also amend the address in the relevant sections of the dossier for the sake of consistency.

2. Update because of tonnage band increase means in this case an update to the highest tonnage band (1000 t/a and above).

    As the substance already had been registered in the tonnage band of 100 t/a to 1000 t/a the registrant already paid the fees for this tonnage band in the past. For the update the registrant will be obliged to pay the difference between the fees for the highest tonnage band and the fees that already had been paid.

    For a registrant who acts as a Member of Joint submission the fees in general are 25% less than for a single submission.

    Fees to be paid for the update in accordance with "COMMISSION IMPLEMENTING REGULATION (EU) No 254/2013 of 20 March 2013 amending Regulation (EC) No 340/2008 on the fees and charges payable to the European Chemicals Agency pursuant to Regulation (EC) No 1907/2006 of the European Parliament and of the Council on the Registration, Evaluation, Authorization and Restriction of Chemicals (REACH)":

- Single submission: 33 201 € − 12 317 € = 20 884 €.
- As a Member of Joint submission: 24 901 € − 9237 € = 15 664 €.

3. Transfer of a standard registration into a registration for a transported isolated intermediate will not cause any further fees. The update of substance B can be done without having any further invoice from ECHA.

   For substance A the registrant has to face the fact that he will receive an invoice (for a large company 1714 € as it is a single submission), because for a registration as an on-site isolated intermediate in any case further fees have to be paid. The registrant could only avoid this fee by pretending that he intends to register substance A in the relevant tonnage band as a transported isolated intermediate (including the use as an on-site isolated intermediate).

4. First, ECHA will not reimburse any fees in such a case. The two companies made for whatever reason a stupid mistake by doing a registration in a higher tonnage band than required. As such mistakes may have been done in the past not only by our example companies, here it must be recommended again to think through things carefully before submitting dossiers. If your company starts with manufacturing a new non-phase-in substance you are required to register such a substance before your company is allowed to manufacture 1 t/a and above, but in any case you are allowed to continue with manufacturing once you have done the registration in the smallest tonnage band even when the next threshold is reached. Your company is obliged at the moment there is a tonnage band increase to take further action without undue delay, but as there is no sharp deadline, you may have all the time you need, if you have a proof for having started your further action in time.

   However, we try to solve the above problem for the two companies as we cannot change the past. The Lead registrant is not allowed to do a tonnage downgrade to the tonnage band 1 to 10 t/a as actually the Co-registrant did a registration in the tonnage band 10 to 100 t/a. Therefore, first the Co-registrant has to submit a dossier to ECHA and indicate the tonnage downgrade to the smaller tonnage band. In any case, ECHA will need proof from this company that they never were obliged to have a registration for the tonnage band above 10 t/a − otherwise ECHA never will accept the tonnage downgrade.

   If the Co-registrant was able to do the tonnage downgrade successfully the Lead registrant may do the same. Also in this case ECHA needs proof concerning the volumes manufactured by this registrant in the past. If the Lead registrant can explain the situation in a credible way, ECHA may accept it.

5. No. See also page 15, Section 2.2.5 of the "Guidance on Scientific Research and Development (SR&D) and Product and Process Oriented Research and Development (PPORD), (Guidance for the implementation of REACH), February 2008, published by the European Chemicals Agency".

6. Yes. Update because of change in status or identity of the registrant as described in Article 22 (1) (a) of the REACH regulation, if it concerns a change in the identity of the registrant, for example, because the company was sold or the name of the registrant's company changed, then the update must not be done separately at the dossier level for all substances that the registrant registered. It will be enough to indicate this sort of change by updating the information in REACH-IT.

7.

- The reason for the update: "update on request" or "spontaneous update".
- Submission number of the last successful submission (see also Chapter 10 of this book).

Chapter 12

1. Yes, as a solvent requires a standard registration it may become subject of authorization if it has intrinsic properties as listed in Article 57 of the REACH regulation.

2. Although substance A meets the criteria listed in Article 57 (a) and therefore may be a candidate for inclusion in Annex XIV, the registrant will never have to do an authorization application based on the REACH regulation as it is actually in force. Whereas other registrants that registered substance A with a standard registration may have to do an authorization application in future, if substance A is included in Annex XIV, a registrant that did a registration only for an on-site isolated intermediate or a transported isolated intermediate will benefit from an exemption in accordance with Article 2 (8) (b) of the REACH regulation.

   Article 2 (8) says: "On-site isolated intermediates and transported isolated intermediates shall be exempted from:" . . . "(b) Title VII". As Title VII deals with "AUTHORIZATION" this means simply exemption of on-site isolated intermediates and transported isolated intermediates from any obligation to prepare an authorization application.

3. In general, a substance that is used as an on-site isolated intermediate will not be subject to any restriction nor subject of authorization. Therefore, company A will not have to face any restrictions by using the substance of concern solely as an on-site isolated intermediate.

   For company B the inclusion of this substance in Annex XVII may have consequences. In case of doubt the description in Annex XVII should be studied in detail.

4. As the use of a substance as a transported isolated intermediate and also as an on-site isolated intermediate is not subject of the authorization process, these uses should be omitted.

Chapter 13

1. Substance Volume Tracking for an EU manufacturer in any case is based on the yearly manufactured amount of a certain substance, whereas in the case of a Non-EU manufacturer in principle only volumes imported into the EU per calendar year have to be considered. Volumes supplied to Non-EU customers by a Non-EU manufacturer need not to be considered. In the case of a Non-EU manufacturer, furthermore, it is possible to cover the whole amount supplied to EU customers under a registration done by the Only Representative of the Non-EU manufacturer, but it can be also considered to make the EU customers register in the role of an importer. For several importers the data requirements for each of the importers may be less than if the whole amount supplied to EU customers has to be covered with the registration of the Non-EU manufacturer via an Only Representative. Therefore, a Non-EU manufacturer may have a higher work-load in doing the Substance Volume Tracking, but on the other hand he and his EU customers may also benefit from several advantages in case the total volume supplied to EU customers can be split in several registrations or if a customer purchases less than 1 t/a these amounts can be exempted from registration obligations and hence there will be no restrictions concerning the use of this substance, whereas the EU manufacturer has to take care of the registration obligations as soon as he manufactures 1 t/a and above of a certain substance.

2. see also Chapter 13, Figure 13.7 in this book.
   - Read the description of use in the first part of the exposure scenario.
   - → If the description of use is very different from the way you use the product, you should contact your supplier and discuss it.
   - Compare the information given in the exposure scenario on how the substance or preparation may be used with your own use(s).
   - → If you use the substance or preparation in a way that leads to higher exposure, for example, if you use it more often, in larger amounts or in a different way from that described, you may not comply with the exposure scenario and you should contact your supplier.
   - Compare the risk management measures (RMM) specified in the exposure scenario to the way in which you protect workers, consumers or the environment.

→ Decide if your measures are as, or even more efficient than these recommended in the exposure scenario.

→ You should also inform your supplier if you think the risk management measures he recommends are inappropriate.

- If your use of the substance or preparation differ from the exposure scenario, it may pose risks to your workers, consumers or the environment, therefore further steps must be defined.

→ There are a number of options:

- Contact your supplier and ask him to prepare an exposure scenario that fits your use conditions.
- Change your working practices.
- Assess in more detail if there is actually a risk or not.
- Look for less hazardous substances or preparations that can be used alternatively.

3. In general, only cases where a substance as such will be forwarded to EU customers needs to be considered. This may occur if a Non-EU company purchases for example, a solvent that is used in a manufacturing process and afterwards a certain product is supplied to EU customers as a solution in this solvent. As there are in general registration obligations as soon as there is 1 t/a and above of the solvent supplied to a EU customer, this EU customer may only benefit from being a Downstream User in cases where the Non-EU manufacturer bought the solvent from a European source and can prove that the substance as such is re-imported into the EU.

4.

a) None.

b) As under REACH only business with EU companies involved needs to be considered, supplies to the customer in the USA and also in Switzerland need not be considered. For supplies to the Austrian company, the Swiss company is a Non-EU manufacturer.

c) If the cyclohexane was purchased from an EU source, the Swiss company and also the Austrian customer may be a downstream user for this solvent. If the cyclohexane was not purchased from a (pre-)registered source the Austrian company will have to take care of the registration obligations if the amount of the solvent imported exceeds 1 t/a. Concerning the substance P manufactured at the Swiss company, either the Austrian company may register in the role of the importer or the Swiss company may take care of the registration obligations via an Only Representative.

Appendix – Answers and Solutions Concerning the Sections Examples and Exercises within this Book | 279

5. Besides some duties concerning information in the supply chain there is a certain risk that such a substance will be included in Annex XIV of the REACH regulation and therefore becomes subject of authorization.

If the use of the substance requires a standard registration, it will be necessary to prepare an application for authorization. This may be done either by the supplier or by the user of such a substance.

In any case, an application for authorization will be expensive and there is a certain risk that it may be rejected. If an authorization is granted a certain substance will become more expensive because of the high costs for the authorization. As a consequence, the user will have to increase the price for a product based on such a raw material or cease manufacture because of the high costs.

The situation could also arise that the supplier is not willing to market such a substance any longer and therefore the availability of a certain substance may not be warranted for the future.

The situation may be less risky if the substance is used solely as a transported isolated intermediate under strictly controlled conditions and the registrant therefore did a registration for a transported isolated intermediate, because this use is exempted from authorization.

Chapter 14

1. If a Downstream Supplier (Distributor or Trader) does not use a substance of request, but only forwards this substance to other actors down the supply chain that will use this substance, he will have no obligation to prepare a Downstream User report. If a Downstream User report is required, this has to be done by a User of the substance.

2. All relevant information has to be stored for a period of at least 10 years after the last manufacture, import, supply or use of the substance or preparation. (See Article 36 [7]).

3. ✓ Non-EU manufacturer.

    In accordance with REACH Article 8 (3) it is the obligation of the Non-EU manufacturer to inform the importers within the same supply chain that they can be regarded as Downstream Users.

4. Volumes may not be covered by the registration of the Non-EU manufacturer via Only Representative as the REACH regulation demands only to register the volumes that are sold by a Non-EU manufacturer to his EU customers.

    A Non-EU Trader itself is not allowed to register and he is not allowed to pretend that a certain substance will be covered by the registration of his upstream supplier (the Non-EU manufacturer who registered via Only

Representative) without having the permission of the Non-EU manufacturer. On the other hand, the EU Trader because of the competition law is not allowed to provide the Non-EU Manufacturer with the identity of his EU customers. The EU customers of the Non-EU trader may demand proof of being Downstream Users within the described Supply Chain, but the Only Representative cannot know them, as he is provided only with a list of direct customers by the Non-EU manufacturer.

5. The authorization number.

6. A Downstream User shall notify ECHA within three months of the first supply of the substance.

7. Article 35 of the REACH regulation demands: "Workers and their representatives shall be granted access by their employer to the information provided in accordance with Articles 31 and 32 [8, 9] in relation to substance or preparation that they use or may be exposed to in the course of their work." This means that the employer has to grant access to Safety Data Sheets in accordance with Article 31 [8] and alternatively in the case of substances or preparations for which a Safety Data Sheet is not required there is still the duty to communicate any other information that is detailed in Article 32 [9] – for example, registration number and if so information concerning any restriction or any authorization granted or denied.

References

Chapter 3

1 see Article 3 (1) of the REACH regulation
2 see Article 3 (3) of the REACH regulation
3 see Article 3 (2) of the REACH regulation

Chapter 5

4 see Article 37 (4) of the REACH regulation

Chapter 7

5 see Article 3 (20) (b) of the REACH regulation

6 see Article 22 (1) of the REACH regulation

Chapter 14

7 see Article 36 of the REACH regulation
8 see Article 31 of the REACH regulation
9 see Article 32 of the REACH regulation

Index

Note: Page references in **bold** refer to Figures.

a

actors within a supply chain   231
already notified substances, *see* formerly notified substances
amendment in the Chemical Safety Report   171
application for authorization   190, **191**, 200, **200**
– main elements of   191, **192**
article, definition   27
authorization
– process   181
– uses exempted from   181, 182

c

Candidate Lead Registrant, notification   134
carcinogenic substances   183
carcinogenic, mutagenic or toxic for reproduction (CMR) products, *see* CMRs
change in access to information in the registration   171
change in status or identity of the registrant   167
change in the composition of a substance   168
change of tonnage band   169
Chemical Safety Assessment (CSA)   22, 156
Chemical Safety Report   23, 29, 76, 103, 136, 148, 170, 172
– amendment in   171
China REACH   11
classification, change in   170, 171
CLP regulation   10, 11
CMRs   152, 183, 185, 194
– Annex XV Report for the Identification of a Substance as CMR   187
– harmonized classification and labeling of   186
– registration deadline   211, 218
– socioeconomic assessment (SEA)   196
communication
– by Downstream User or Downstream Supplier   247–249
– – with customers   248
– – from Downstream User with workers   248
– – from Downstream User to authorities   249
– – from Downstream User to ECHA   249
– – with Only Representative of Non-EU manufacturer   248, **248**
– – with suppliers   247, **247**
– by Non-EU distributors or Non-EU traders   243–246
– – with EU customers   244, **244**
– – with Non-EU manufacturers   245, **245**, **246**
– – with Only Representative acting on behalf of Non-EU manufacturer (supplier)   246, **246**
– by Non-EU manufacturers   235–243
– – to EU customer   236, **238**
– – to Non-EU customers   236–241, **239**, **241**, **242**
– – to Only Representative   236, **237**
– – to suppliers   241–243, **243**
– – sample letter to inform customers about Appointment of an Only Representative   **239**
– obligations (REACH regulation)   229–231, **230**

– by Only Representative 249–251
– – with customers of the Non-EU manufacturer 250, 251, **252**
– – with ECHA and National Authorities 250
– – with Non-EU manufacturer 249, 250, **250**
– within Pre-SIEF 124–127
– by suppliers 231–235
– – with authorities 234, 235
– – to EU customers 231, **232**
– – information for workers 233, 234, **234**
– – to Non-EU customers 232, 233, **233**
– – obligations in the context of authorization 235, **235**
– – with Upstream Supplier 234
– in supply chain 229–251
Competition Law 9, 10, 240
confidentiality flags 109
Consortium Agreement 137
Cooperation Agreement 137
costs/fees 103, 104, **104**
– Member of Joint submission 115
– registration dossier updates 175, 176
Custom Manufacturing 43

d
data
– requirements 83, 84, **85**
– sharing 135–140
– sharing disputes 140, 141
Data Holders 124, 125, **125**
data requirements and documentation for an application of authorisation 196–199
Directives
– Commission Directive 2000/21/EC 7
– Commission Directive 91/155/EEC 7
– Commission Directive 93/105/EC 7
– Commission Directive 93/67/EEC 7
– Council Directive 67/548/EEC 7
– Council Directive 76/768/EEC 8, 182
– Council Directive 76/769/EEC 7
– Directive 1999/45/EC 7, 155, 182
– Directive 2001/82/EC of the European Parliament and of the Council of 6 November 2001 39
– Directive 2004/37/EC 8
– Directive 2006/12/EC 37
– Directive 67/548/EEC 1, 32, 33, 36, 42, 53, 55, 80, 107–109, 122, 152, 155, 172–174, 182, 183, 185
– Directive 90/385/EEC 181
– Directive 91/414/EEC 41, 125, 181
– Directive 92/32/EEC 32
– Directive 93/42/EEC 181
– Directive 98/8/EC (Biocidal Product) 41, 125, 181
– Directive 98/70/EC 181
– Directive 98/79/EC 181
documentation
– correct use of TIIs by customers 224, 225, **224**
– from Downstream Users in regard to Article 18(4) 224, 225
– inhouse, in regard to intermediates 224
– manufacturing process of OIIs and TIIs 224
dossier preparation 83–104
– on-site isolated intermediate 91
– PPORD 89, **90**
– standard registration (full registration) 95–98
– – Lead Company/single submission **98**
– – manufacturers, importers and EU Only Representatives Member of Joint submission **99**
– – substance data set preparation in IUCLID5
– – EU Manufacturers and Importers acting as Lead Registrant **96**
– – Members of Joint Submission **97**
– transported isolated intermediate 93–96
– see also inquiry dossier; registration dossier
dossier submission, fees 103, **104**
Downstream Supplier
– communication by 247–249
Downstream Users 16, 20–23, **50**, 61
– checklist for 24
– communication by 247–249
– documentation from 224, 225
– Letter from registrant to, concerning identified uses and template for the attached questionnaire 62–69
– Letter to Supplier to make their uses identified uses 69
– obligations of 158, 159
– raw materials and company as 214, **215**

e
EC numbers, official vs non official 33, 34, **34**
EINECS 1, 3, 32
EINECS number 124
ELOC, identification of 186, 187
European Chemicals Agency (ECHA) 2, 3, 23, 109
– fees and charges payable to 8, 9
European Economic Areas (EEA) 2, 3, 16

exemptions from REACH 34–42
– food and feedingstuffs 39, 40
– medicinal products 39, 40
– non-isolated intermediates 35
– parts falling under REACH and part exempted 42
– polymers 37, 38
– PPORD 40
– Re-imported substances 39
– substances in the interest of defense 36
– substances listed in Annex V 36
– substances manufactured or imported in amounts below 1 t/a 35
– substances mentioned in Annex IV 35
– substances regarded as being registered 40–42
– Waste and recovered substances 37
Exposure Scenario (ES) 156
extended Safety Data Sheets (eSDS) 9–11, 70, 155–157

f
fees, *see* costs/fees
Fleischer list 140
food and feedingstuffs 39, 40
formerly notified substances 33, 80
– regarded as Registered under REACH 107, 108
– confidentiality claims 173, 174
– dossier update 109, 172–174
– Registration Dossier updating of 109, 172–174
– Registration Number claiming under REACH 108
– Registration Number for, claiming for 107–110
– tonnage band increase 172, 173
full registration 70

g
Globally Harmonized System of Classification and Labeling of Chemicals (GHS) 10, 11

h
history of REACH 1–3

i
Importer 16, 49, **50**
initial verification after submission of a registration dossier 147, **148**
inquiry dossier 90, 91, **92**
– non-phase-in substances 111–113

intermediate substances 30, 31
– *see also* non-isolated intermediates; on-site isolated intermediate (OIIs); transported isolated intermediates (TIIs)
IUCLID (International Uniform Chemical Information Database) 84–88, **86–88**
– Substance Data set 113
IUCLID5.4, entering data and information in 98–103
– IUCLID Section 1.2 98, 99
– IUCLID Section 1.3 99, 100
– IUCLID Section 1.4 100
– IUCLID Section 1.7 101
– IUCLID Section 2.3 101, 102
– IUCLID Section 3.1 102
– IUCLID Chapter 11 102, 103
– IUCLID Chapter 13 103
IUPAC name 124

j
Joint Submission 115, 116

k
Korea REACH 11

l
labeling, change in 170, 171
late pre-registration 53–55, **54**, 121–124
Lead Company **113**, 115, 128, 129, 136, **170**
Lead Registrant 115, 142, **170**
– Agreement **130**, 131–134, **133**
– defining **132**
– notification 134
Letter of Access (LoA) 137–140
– concerning data as studies and tests 138, 139
– to a registration dossier 139, 140

m
manufacturer 15, 16, 49, **50**
– within EU 15, 16
– non-EU 16–20
medicinal products 39, 40
Member of Joint submission, Registration as 115
Member State Competent Authority 108
mono-constituent substances 28
monomer 37
monomer unit 38
multi-constituent substances 28, 29
mutagenic substances 183

n

nitrogen   35
new identified uses   169
no-longer polymers (NLP)   32, 33
non-isolated intermediate   30, 35
non-phase-in substances   32, 33, 107
– checklist for Business Managers   **43, 44**
– inquiry dossier   111–113
– – identity of inquirer   112
– – submission previous to registration   112
– – time frame   113
– phase-in substance vs   112
– REACH vs needs of industry acting in the market   **117**
– registration of   111–118
– – difficulties and problems   117, 118
– registration dossier preparation   113–115
– – joint submission with other potential registrant(s)   115, 116
– – Member of Joint Submission   115
– – outcome of inquiry   113, **114**
– – single submission   116
notified new substances (NONS)   1, 107
Notified Substances   41

o

obligation to submit registration dossier   49–60
– amounts to be considered by importers   **58**
– globally acting enterprises/decision-making process   59
– REACH relevance for substances manufactured within/outside EU   **57**
– special rules for Non-EU manufacturers   56–59
Only Representative (OR)   16, 17, 20, 22, 25, 49, **50**, 51, 58, 76, 79, **108**
On-site isolated intermediate (OII)   30, 31
– dossier   91, 92, **93**
– documentation of manufacturing process of   224, 225, **224**
– registration as   70, 71, **93**, 94
opt-out   142, 143, **144**
– key messages   **144**
overall completeness check   149, 150
oxygen   35

p

PBT substances   186, 188
phase-in substances   31, 32
– pre-registration and late pre-registration, preparation for   121–124
– – format of (late) pre-registration number   **124**
– – timelines for   **122**
– registration process   121–145
plant-protection products   41
polymers   37, 38, 70
Potential Registrants   124, **125**
PPORD (product and process oriented research and development)   8, 80
– dossier   89, **90**
– exemption from REACH   40
– notification   100
– notification fees   103, **104**
preparation, definition   27
pre-registration   53–55, **54**, 121–124
pre-registration number   **124**
pre-SIEF, communication within   124–127
– data holders   125, **125**
– duties and rights of SIEF participants   126, 127
– potential registrants   126, **127**
– Third Party Representatives (TPR)   126
pre-SIEF Survey   **130**, **133**
purity, substance   27, 28
purpose and scope of REACH   4, 5

q

quality observation letter (QOBL)   165

r

raw materials, list of used   207
– company as Downstream User   214, **215**
– parties involved in ensuring compliance   **208**
– process after receiving a SDS or an eSDS from supplier   216, 217, **217**
– – information forwarded to customers down the supply chain   216, 217
– – safety data sheets and exposure scenarios checks   216, **217**
– registration deadline based on properties   211, **212**
– role definition under REACH   208–211, **209, 210**
– uses of certain substances within company   211–214, **213**
REACH-IT   34, 50, 147–149, 151, 166, 168
– company UUID   134
– data holders in   125
– Inquiry Dossier preparation   88
– Member of Joint Sumission in   115
– overall completeness check in   149
– potential registrants and   126

– pre-registration in   53, 54, 122, 124
– pre-SIEF members in   124, 128, 129, 132
– reference number   150
– registration number in   107, 108
– SFF role in   128, 132
– SIEF and   127
– submission of application of authorization   200
REACH Registration Certificate   **253**
Registrant, obligations of   155–158
registration as a substance   70
registration deadline   55, 56
– for non-phase-in substances   **55**
– for (late) pre-registered substance   **55**
registration dossier
– of Lead Company   144
– of Members of Joint Submission   144
– non-phase-in substances   113–116
– obligation to submit   49–60
– after submission to ECHA   147–159
registration number
– for formerly notified substances, claiming for   107–110
registration
– types of   103
Regulations
– Commission Regulation (EC) No 2032/2003   41
– Council Regulation (EC) No 2494/95   9
– Council Regulation (EEC) No 793/93   7
– Regulation (EU) No 254/2013   175, 200
– Regulation (EC) No 340/2008   8, 9, 175
– Regulation (EC) No 703/2001   41
– Regulation (EC) No 726/2004   6, 39
– Regulation (EC) No 178/2002   40
– Regulation (EC) No 1272/2008   1
– Regulation (EC) No 1488/94   7
– Regulation (EC) No 1490/2002   41
– Regulation (EC) No 1935/2004   182
– Regulation (EEC) No 3600/92   41
Re-imported substances   39
Requested Update   202

S

Safety Data Sheets   11, 70, 155–157
Substance Identification Profile (SIP)   130, **131**
Substance Sameness   130
Sameness Check   9, 112
scope of REACH   4, 5
SDS, *see* Safety Data Sheets
SIEF   104
– agreement   137
– cooperation within   134

– obligations of participants   135
– participants   126, 127, **127**
SIEF formation   127–134
– Lead Registrant Agreement   **130**, 131–134, **133**
– Lead Registrant, definition   **132**
– Lead Registrant Notification   134
– pre-SIEF Survey   **130**, **133**
– substance sameness and Substance Identification Profile (SIP)   130, 131, **131**
SIEF Formation Facilitator (SFF)   127–129, **129**, 132
– tasks of   **129**
SNIF file   108, 109
Sole Representative, role of   107, 108
spontaneous update   163, **164**, 202
standard registration   70
status or identity of the registrant, change in   167, 168, **167**
after submission of Registration Dossier to ECHA   147–159
– dossier evaluation   152–155
– – compliance check of registration   153, 154
– – examination of testing proposals   152, 153
– – on-site isolated intermediates   155
– – substance evaluation   154, 155
– – time periods for examination of testing proposals/draft decision   **153**
– end of pipeline activities   150–152
– fees   149
– format of submission number   **148**
– initial verification   147, 148, **148**
– obligation to update information   158
– obligations of Downstream Users   158, 159
– obligations of Registrant   155–158
– overall completeness check   149, 150
– reference numbers   150, **151**
– technical-completeness-check-plug-in (TCC plug-in)   147–149
substance
– composition   28
– composition, change in   168, 169
– definition   27
– mono-constituent   28
– multi-constituent   28, 29
– UVCB   29
– types of use   29–31
– with end use   29
Substance Identification Profile (SIP)   130, 131, **131**
Substance Information Profile   9

substances manufactured, list of   217–224
– identification of registration obligations   218, **219**, **220**
– registration deadline   218–221
– uses at company's own site and of customers   221–224, **223**
Substances of Very High Concern (SVHC)   181–203
– application for authorization   190, **191**, 200, **200**
– – adequate control route   193, 194–196
– – analysis of alternatives   **194**
– – main elements   191–196, **192**
– – preparation of a substitution plan   **195**
– – socioeconomic (SEA) route   193, 196
– – data requirements and documentation   196–199
– – identity and composition depending on type of substance   197
– – information in dossier header   199
– – IUCLID Chapter 13   199
– – IUCLID Section 1.3   198, **198**
– – IUCLID Section 3.5   198
– – IUCLID Sections 1.1 and 1.2   197
– – relevant IUCLID chapters   **197**
– deadlines   200
– fees   200, **200**
– To Dos after granting of an authorization   202
– information in Annex XIV   188, 189
– prioritization and inclusion in Annex XIV   188
– To Dos after refusal of an authorization   202, 203
– requested update   202
– restrictions and information in Annex XVII   189, 190
– review of authorizations   203
– spontaneous update   202
– submission of the application for authorization   201–203
– subsequent applicants and their obligations   201
– substance identification and identification procedure   184–187
– – Annex XV dossier   185–187, **186**
– – Annex XV report   187
– – content of an Annex XV dossier   185, 186, **186**
– – preparation of an Annex XV dossier   186
– – procedure   185

– uses exempted from authorization   181, 182, **183**
– Candidate List   187, 188
Substance Volume Tracking (SVT)   **21**, 157, 158, 169, 225–227
– for EU Manufacturer   225, **226**
– for Non-EU Manufacturer   225, **227**

t

Table of Contents of REACH   3, 4, **5**, **6**
Technical-Completeness-Check-Plug-in (TCC Plug-in)   147–149
Third Party Representatives (TPR)   124, **125**, 126
toluene   190
tonnage band for registration   55, 56
– change in   169
toxic to reproduction   183
Trader   **50**
– EU vs non-EU   23–25
Transported Isolated Intermediates (TIIs)   31, 157
– dossier   92–96
– registration of   71, **71–80**
– correct use of, by Customers   224, 225, 224
– manufacturing process of   224, 225, 224
– letter to ask for confirmation concerning Article 18 (4)   71, **72**
– letter to Non-EU customers unwilling to sign Article 18 (4) confirmation   72, **73**, 74
– checklist for manufacture and strictly controlled conditions   **75**, **76**
– format for documenting inhouse information on strictly controlled conditions   **77**
– format for documentation information on risk management   **78**, **79**

u

United Nations (UN)   10
UVCB (Substances of Unknown or Variable composition, Complex reaction products or Biological materials)   28, 29
update of registration dossier   163–176
– access granted to information in the registration, change in   171
– Chemical Safety Report amendment   171
– classification and labeling, change in   170, 171
– costs/fees   175, 176

- formerly notified substances 109, 172–174
-- because of tonnage band increase 172, 173
-- confidentiality claims 173, 174
- further tests 171
-- compliance check 165
-- dossier evaluation 165
-- missing information 164
- new identified uses 169, 170
- PPORD notifications 174, 175, **175**
- reasons for **164**
- requested 164–166
-- after examination of testing proposals 166
- spontaneous 163, 166, 167, **167**
- status or identity of the registrant, change in 167, 168
- substance composition, change in 168, 169
- substance risks, new knowledge of 170
- timing 163
- tonnage band, change in 169
- update on request 163
Use Descriptor system 61
UUID 134

v

vertebrate animals 112
- testing 166
vPvB substances 186, 188

w

waste and recovered substances 37
well-defined substances 28